Springer Series in Materials Science

Volume 333

The Springer Series in Materials Science covers the complete spectrum of materials research and technology, including fundamental principles, physical properties, materials theory and design. Recognizing the increasing importance of materials science in future device technologies, the book titles in this series reflect the state-of-the-art in understanding and controlling the structure and properties of all important classes of materials.

Lorenz Holzer · Philip Marmet · Mathias Fingerle ·
Andreas Wiegmann · Matthias Neumann ·
Volker Schmidt

Tortuosity and Microstructure Effects in Porous Media

Classical Theories, Empirical Data and Modern Methods

 Springer

Lorenz Holzer
School of Engineering, Institute
of Computational Physics
Zurich University of Applied Sciences
Winterthur, Switzerland

Philip Marmet
School of Engineering, Institute
of Computational Physics
Zurich University of Applied Sciences
Winterthur, Switzerland

Mathias Fingerle
Math2Market GmbH
Kaiserslautern, Germany

Andreas Wiegmann
Math2Market GmbH
Kaiserslautern, Germany

Matthias Neumann
Institute of Stochastics
Ulm University
Ulm, Germany

Volker Schmidt
Institute of Stochastics
Ulm University
Ulm, Germany

Publiziert mit Unterstützung des Schweizerischen Nationalfonds zur Förderung der wissenschaftlichen Forschung. Die Druckvorstufe dieser Publikation wurde vom Schweizerischen Nationalfonds zur Förderung der wissenschaftlichen Forschung unterstützt.

ISSN 0933-033X   ISSN 2196-2812 (electronic)
Springer Series in Materials Science
ISBN 978-3-031-30479-8   ISBN 978-3-031-30477-4 (eBook)
https://doi.org/10.1007/978-3-031-30477-4

This Springer imprint is published by the registered company Springer Nature Switzerland AG
The registered company address is: Gewerbestrasse 11, 6330 Cham, Switzerland

Preface

Tortuosity is an important morphological characteristic, which describes the limiting effects of the pore structure on the transport properties of porous media. In this book, the relevant aspects of tortuosity and associated transport in porous media are reviewed thoroughly.

The *classical theories, definitions and concepts related to tortuosity* and associated equations for the prediction of effective transport properties are summarized separately for flow, conduction and diffusion. These theories and underlying concepts evolved over a long period, and their evolution is tightly linked with the progress of relevant methodologies such as tomography and 3D image analysis. As a result of this long history, many different definitions and nomenclatures can be found in literature, which is the source of severe confusion and frequent misconception. In order to clarify the discussion on this topic, *a new classification scheme and a systematic nomenclature for the different tortuosity types are proposed*. Three main classes of tortuosity are distinguished: (a) direct geometric, (b) indirect physics-based and (c) mixed (geometric and physics-based) tortuosities.

An extensive *review of empirical data focusing on tortuosity–porosity relationships* reveals a *systematic pattern associated with the different tortuosity types*. For example, the values of direct geometric as well as mixed tortuosities are systematically lower than those from indirect physics-based tortuosities. Systematic differences can be observed even within single tortuosity classes. For example, when comparing different types of direct geometric tortuosities, the values for medial axis tortuosity are systematically larger than those for geodesic tortuosity. Hence, a scientific treatment of tortuosity always has to provide a clear definition of the tortuosity type under consideration, which is tightly related to the underlying method of computation.

The *review of methods for characterization of porous media, in general, and computation of tortuosity, in specific*, includes the following disciplines: 3D imaging, image processing (qualitative and quantitative), numerical transport simulation, stochastic geometry and virtual materials testing. Strong emphasis is put on the description of the methods of computation and calculation, which are specific for the different types of tortuosities. An extensive *list of available software packages* with their modular options is also provided.

Finally, *mathematical and empirical expressions for microstructure–property relationships* are discussed for (a) conduction and diffusion and (b) flow/permeability. The evolution of these expressions is intimately related to the methodological improvements in tomography, image processing, numerical simulation, stochastic geometry and virtual materials testing. This evolution led to a better understanding of the different tortuosity types and of their impact on transport properties. However, the methodological progress also resulted in the perception of *additional relevant characteristics such as the constrictivity and, in case of viscous flow, the hydraulic radius*. Hence, new mathematical expressions for microstructure–property relationships make use of modern methods that enable to characterize these morphological characteristics in a specific way (i.e., specific computation of tortuosity type, constrictivity and/or hydraulic radius). Compared to classical equations (e.g., the Carman–Kozeny equation or Archie's law), the *new expressions provide a higher prediction power of effective properties, in particular for porous and composite materials with complex microstructures.*

Winterthur, Switzerland   Lorenz Holzer
Winterthur, Switzerland   Philip Marmet
Kaiserslautern, Germany   Mathias Fingerle
Kaiserslautern, Germany   Andreas Wiegmann
Ulm, Germany   Matthias Neumann
Ulm, Germany   Volker Schmidt

Acknowledgements

This book arose from the work in numerous projects dealing with microstructure analysis of porous media and with microstructure optimization of engineered materials. The most recent projects to mention in this context are '*Volta*' (funded by Swiss Federal Office of Energy-SFOE, Grant Nr SI/501792-01—8100076), '*GeoCloud*' (funded by Eurostars/EU/Horizon Europe, Grant Nr E!115455), '*Hikomat*' (funded by the German Federal Ministry for Economic Affairs and Energy BMWi granted through Project Management Jülich, Grant Nr 03ET6095E) and '*HiStructures*' (funded by the German Federal Ministry of Education and Research BMBF, Grant Nr 03XP0243D). Moreover, this book contributes to the research performed at *CELEST* (Center for Electrochemical Energy Storage Ulm-Karlsruhe). The work by MN was funded by the German Research Foundation (DFG) under Project ID 390874152 (*POLiS* Cluster of Excellence, EXC 2154).

These financial supports are gratefully acknowledged. We would also like to thank all the project partners, researchers and students, that contributed to this book with many fruitful discussions, with critical questions and with constructive comments.

Open access to this book is published with support from Swiss National Science Foundation (SNSF, Nr. 10BP12_215232/1).

Contents

Chapter 1
Introduction

Tortuosity (τ) is widely recognized as a key concept for transport in porous media, which describes the impact of pore structure on the effective transport properties. Tortuosity is a dimensionless parameter that depends on the windedness of transport pathways. The increase of path lengths due to tortuous pore morphology can contribute significantly to the transport resistance. This concept was introduced almost 100 years ago by Kozeny in 1927 [1]. Since then, tortuosity has been widely applied in various research disciplines, such as chemical engineering, geoscience, materials science, and life science. Uncountable studies dealing with tortuosity have been performed, using different methodologies (physical theory, laboratory experiments, numerical simulations, 3D imaging and image processing) and combinations thereof. It is beyond the scope of this book to present a detailed review of all these studies. For this purpose, we refer to excellent review articles on tortuosity in specific (e.g., by Clennell [2], Ghanbarian et al. [3], Shen and Chen [4], Tjaden et al. [5]) and on transport in porous media in general (e.g., [6–15]).

In order to explain the focus of this paper, it must be emphasized that until now there exists no unifying theory for the tortuosity concept. Therefore, the discussion of tortuosity bears considerable potential for confusion. Many different definitions of tortuosity have been presented, depending either on the characterization method (direct geometric 3D analysis by tomography and image processing vs. indirect calculation from effective properties) and/or on the underlying transport mechanism (flow, diffusion and conduction). Confusion is amplified by the fact that many different tortuosity-terms are in use (see Table 1.1). Unfortunately, in many cases there exists no clear definition for these terms and moreover, a globally accepted classification scheme as well as a systematic nomenclature for the different tortuosity types are missing.

As will be discussed in this book, the different tortuosity terms have distinct meanings and can therefore not be used interchangeably. However, the meaning of a specific tortuosity term is often strongly related to the methodology by which tortuosity is determined. Therefore, the topic of tortuosity must be discussed in

© The Author(s) 2023
L. Holzer et al., *Tortuosity and Microstructure Effects in Porous Media*,
Springer Series in Materials Science 333,
https://doi.org/10.1007/978-3-031-30477-4_1

Table 1.1 List of tortuosity (τ) terms from literature

• Geometric τ	• Fudge factor τ
• Geodesic τ	• Retardation factor τ
• Medial axis τ	• Streamline τ
• Percolation path τ	• Volume averaged τ
• Fast marching method (FMM) τ	• Area averaged τ
• Distance propagating method (DPM) τ	• Path-length τ
• Pore centroid τ	• Random walk τ
• Path tracking method (PTM) τ	• Relative τ
• Pore throat τ	• Formation τ
• (Bulk) diffusional τ	• Fractal τ
• Knudsen τ	• (In)active phase τ
• Electric τ	• Total electrode τ
• Ionic τ	• Characteristic τ
• Thermal τ	• Experimental τ
• Hydraulic τ	• Impedance (EIS) τ
• Flux-based/physics-based τ	• Crack τ
• Indirect or inverse τ	• Three phase boundary (TPB) τ
• Direct τ	• τ Factor (T)
• Kinematic τ	• τ Tensor

context with the corresponding methodologies (i.e., 3D imaging and image analysis, transport simulation or laboratory experiments) and their continuing development.

Initially, the basic theories on tortuosity (e.g., Carman-Kozeny equations) were developed at a time when direct measurement of tortuosity by means of tomography and 3D image analysis was not possible. Therefore, tortuosity was determined indirectly—usually from effective transport properties that were measured experimentally. This led to a certain gap between theoretical descriptions, which are based on considerations of path lengths in simplified geometric models (e.g., in bundles of tubes or in packed spheres), and empirical investigations, which derive tortuosity values indirectly from bulk effective properties. Hence, different definitions for tortuosity evolved over time, depending on the basic approach (theory vs. experiment vs. modeling), depending also on the field of research and on an associated 'school of thinking' (e.g., petro-physics vs. electrochemistry), and depending also largely on the availability of certain characterization techniques (e.g., computational methods for pore scale modeling or techniques for 3D analysis by tomography and image processing).

Over the last two decades significant progress was achieved in high-resolution tomography as well as in stochastic modeling and numerical simulation of 3D image data representing the morphology of microstructures. These methodological improvements open new possibilities for studying microstructure-property relationships, in general, as well as for measuring tortuosity directly from the microstructure by means of 3D analysis and transport simulation. Due to the availability of new methods, it is now possible to compare different tortuosity concepts and establish correlations between the different tortuosity types. These new possibilities are the basis for the present review, which is structured as follows:

In Chap. 2, *the classical theories and concepts of tortuosity* (starting with the Carman-Kozeny equations), as well as the underlying definitions for the most important tortuosity types are presented in a chronological (historical) order. At the end of Chap. 2, a new classification-scheme is introduced together with a systematic tortuosity-nomenclature. Three main categories are distinguished: direct geometric tortuosities, indirect physics-based tortuosities and mixed (i.e., geometric and physics-based) tortuosities. This classification may help to avoid confusion in future debates.

In Chap. 3, *empirical data from literature* is collected and compared. The collection includes more than 2000 data-points (i.e., tortuosity-porosity-couples) from 70 studies in various fields such as geology, battery and fuel cell research. Thereby, experimental approaches are considered as well as investigations that are based on computational modeling and simulation. The collection of literature data represents the basis for an empirical description, which shows how tortuosity varies for different types of materials and microstructures. Furthermore, in many of these studies different types of tortuosities are measured for the same materials. These datasets enable to define a relative order among the different tortuosity types. More precisely, it turns out that for a given material, the values of certain types of tortuosities tend to be systematically lower than the values of other tortuosity types. This comparison of different tortuosity types results in a surprisingly clear and consistent pattern.

In Chap. 4, *modern methods for microstructure characterization* and associated *calculation approaches for tortuosity* are reviewed. Chapter 4 is structured according to the workflow, which is typical for this kind of microstructure characterization. First, an overview of modern tomography methods is presented with a special emphasis on recent innovations and on current trends. Subsequently, calculation approaches by image analysis and by transport simulation are discussed for all three tortuosity categories: direct geometric, indirect physics-based and mixed tortuosities. In addition, an extensive list with available software packages for image processing, which include codes for the computation of specific tortuosity types, is presented. Finally, modern methods of stochastic geometry used for virtual materials testing are discussed in context with their applications in Digital Materials Design (DMD) and Digital Rock Physics (DRP), which are all strongly associated with the investigation of tortuosity.

In Chap. 5, it is discussed how the recent progress in tomography, 3D image analysis, microstructure modeling and virtual materials testing can be used for a thorough understanding of *microstructure-property relationships*. Based on modern 3D characterization techniques, the effects from tortuous pathways can now be distinguished from other microstructure effects, such as the limitations arising from narrow bottlenecks and from the friction at pore walls. New morphological descriptors were introduced for the bottleneck effect (i.e., constrictivity), for the wall friction effect (i.e., hydraulic radius) and also for the path length effect (i.e., tortuosity). Consequently, new formulas describing the complex relationships between microstructure and effective transport properties have been established recently. The evolution of morphological descriptors and associated formulas describing the micro–macro relationships are reviewed in Chap. 5. For porous media with random microstructures, these new formulas have a higher prediction power compared to traditional equations from the literature, such as e.g., the Carman-Kozeny equation for viscous flow.

Finally, a *summary and conclusions* are presented in Chap. 6.

References

1. J. Kozeny, Über Kapillare Leitung Des Wassers Im Boden. Sitzungsbericht Der Akademie Der Wissenschaften Wien **136**, 271 (1927)
2. M.B. Clennell, Tortuosity: a guide through the maze, in Developments in Petrophysics, ed by M.A. Lovell, P.K. Harvey (Geol. Soc. Spec. Publ. No. 122, 1997), pp. 299–344
3. B. Ghanbarian, A.G. Hunt, R.P. Ewing, M. Sahimi, Tortuosity in porous media: a critical review. Soil Sci. Soc. Am. J. **77**, 1461 (2013)
4. L. Shen, Z. Chen, Critical review of the impact of tortuosity on diffusion. Chem. Eng. Sci. **62**, 3748 (2007)
5. B. Tjaden, D.J.L. Brett, P.R. Shearing, Tortuosity in electrochemical devices: a review of calculation approaches. Int. Mater. Rev. **63**, 47 (2018)
6. P.M. Adler, *Porous Media: Geometry and Transports, Series in Chemical Engineering* (Butterworth-Heinemann, Boston, 2013)
7. J. Bear, *Dynamics of Fluid in Porous Media* (New York, 1972)
8. R.B. Bird, W.E. Steward, E.N. Lightfood, *Transport Phenomena*, 2nd edn. (John Wiley & Sons, New York, 2007)
9. F.A.L. Dullien, *Porous Media: Fluid Transport and Pore Structure* (Academic Press Ltd., London, 2012)
10. R. Hilfier, Local-porosity theory for flow in porous media. Phys. Rev. B **B45**, 7115 (1992)
11. R. Hilfier, Geometric and dielectric chracterization of porous media. Phys. Rev. B **44**, 60 (1991)
12. A.E. Scheidegger, *The Physics of Flow through Porous Media*, 3rd edn. (Univ. of Toronto Press, Toronto, 1974)
13. S. Torquato, *Random Heterogeneous Materials: Microstructure and Macroscopic Properties* (Springer, New York, 2002)
14. F. Willot, S. Forest, *Physics and Mechanics of Random Media: From Morphology to Material Properties* (Presses des MINES, Paris, 2018)
15. J. Bear, *Modeling Phenomena of Flow and Transport in Porous Media*, vol. 1 (Springer International Publishing, Cham, 2018)

Chapter 2
Review of Theories and a New Classification of Tortuosity Types

Abstract Many different definitions of tortuosity can be found in literature. In addition, also many different methodologies are nowadays available to measure or to calculate tortuosity. This leads to confusion and misunderstanding in scientific discussions of the topic. In this chapter, a thorough review of all relevant tortuosity types is presented. Thereby, the underlying concepts, definitions and associated theories are discussed in detail and for each tortuosity type separately. In total, more than 20 different tortuosity types are distinguished in this chapter. In order to avoid misinterpretation of scientific data and misunderstandings in scientific discussions, we introduce a new classification scheme for tortuosity, as well as a systematic nomenclature, which helps to address the inherent differences in a clear and efficient way. Basically, all relevant tortuosity types can be grouped into three main categories, which are (a) the indirect physics-based tortuosities, (b) the direct geometric tortuosities and (c) the mixed tortuosities. Significant differences among these tortuosity types are detected, when applying the different methods and concepts to the same material or microstructure. The present review of the involved tortuosity concepts shall serve as a basis for a better understanding of the inherent differences. The proposed classification and nomenclature shall contribute to more precise and unequivocal descriptions of tortuosity.

Remark Although tortuosity is related to porous media transport, it must be emphasized that the purpose of this chapter is not to give a review on transport equations and associated simulation of transport in porous media. For such topics we refer to dedicated books (e.g., Bird et al. [1]).

2.1 Introduction

Over the last 100 years, many different approaches were developed, how tortuosity can be defined and measured. A unifying concept for tortuosity is still lacking and therefore it is not easy to understand the difference between these numerous existing

© The Author(s) 2023
L. Holzer et al., *Tortuosity and Microstructure Effects in Porous Media*,
Springer Series in Materials Science 333,
https://doi.org/10.1007/978-3-031-30477-4_2

tortuosity types. Furthermore, there exists no suitable nomenclature that helps to address specific tortuosity types in a clear and simple way. In this chapter, the classical concepts and theories of all relevant tortuosity types are reviewed, and a new nomenclature is introduced.

2.1.1 Basic Concept of Tortuosity

For a given porous medium, tortuosity (τ) is basically defined as the ratio of effective path length (L_{eff}) over direct path length (L_0) through the considered porous medium ([2–5]), i.e.

$$\tau = \frac{L_{eff}}{L_0}.$$ (2.1)

Figure 2.1 schematically illustrates L_{eff} and L_0. Thereby, the direct path length (L_0) is easily captured, since it is the sample dimension in transport direction. For theoretical treatment of tortuosity, one then simply has to find a suitable definition of the effective path length (L_{eff}), and for practical application one simply has to find a suitable method to measure effective path length.

2.1.2 Basic Challenges

Unfortunately, definition and measurement of the effective path length (L_{eff}) are not as simple as it appears at a first glance, which explains the emergence of numerous different tortuosity concepts over time. As will be shown in Chap. 3, the different types of tortuosities often reveal significantly different values when applied to the same microstructure. These differences and the underlying proliferation of concepts are usually not properly accounted for in the description of tortuosity, e.g., in appropriate papers and conference presentations. This frequently deficient description of tortuosity is partly caused by the fact that there exists no suitable nomenclature for the different types of tortuosities. In addition, very often, researchers in this field are not aware of the inherent differences between the various tortuosity types. This lack of awareness often becomes the source of confusion in scientific debate, and it can also be the source of misinterpretation of acquired data. The basic challenges in this field are thus to foster the *awareness in the scientific community for the systematic differences* between tortuosity types, and to *introduce a useful classification scheme and nomenclature* that can then serve as a basis for more precise descriptions and for clearer scientific discussions of the topic.

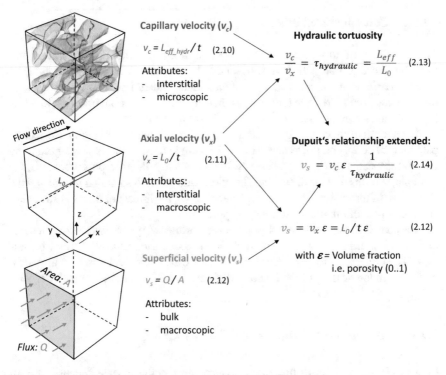

Fig. 2.1 Illustration of three different concepts of hydraulic flow velocity

2.1.3   Criteria for Classification

Two main criteria will be used to classify the different tortuosity types:

2.1.3.1   Method of Determination

Initially, there were no suitable methods available for direct measurement of effective path lengths (L_{eff}) from the microstructure. Thus, for a long period, tortuosity was calculated *indirectly*, using information such as effective transport property and porosity. Over the last two decades, new methods for 3D analysis became available, including micro- and nano-tomography, 3D image processing and numerical simulation. Nowadays, these methods enable us to measure tortuosity and effective path lengths (L_{eff}), respectively, *directly* from the 3D microstructure. Researchers can now choose from a multitude of direct and indirect methods. Understanding the underlying systematic differences between those tortuosity types is crucial, for example, to make a sensible choice of methods and concepts when planning an investigation of porous media.

2.1.3.2 Concept of Definition

The definition of tortuosity can be approached from different viewpoints. For example, the effective path length (L_{eff}) can be considered as a purely *geometric characteristic* of the microstructure, which is independent from the involved transport process. The geometric tortuosity is thus an intrinsic material property, and it is typically determined with quantitative 3D image analysis.

An alternative viewpoint for the definition of tortuosity and description of effective path length focuses on the tracking of a microscopic particle on its way through the porous material. Thereby, it makes a difference whether the transport process is, for example, viscous flow or diffusion. This viewpoint leads to so-called *physics-based definitions* of tortuosity, which are considering both, the impact of material structure, and the impact of the involved transport process.

Nowadays, a multitude of physics-based tortuosity types (hydraulic, electric, diffusional, thermal) as well as many different geometric tortuosity types (medial-axis, skeleton, geodesic, percolation path etc.) are available. Our aim is to provide a profound understanding of the inherent differences between these tortuosity types.

2.1.4 Content and Structure of This Chapter

In this chapter, we describe the underlying concepts, definitions, and theories for all relevant tortuosity types. Thereby, the evolution of emerging definitions and concepts is presented in a chronological (historical) order. The concept of tortuosity was initially introduced in context with the Carman-Kozeny equations for flow in porous media. All other tortuosity types then evolved and diverged from there. Different branches of physics-based tortuosities (i.e., hydraulic, electric, and diffusional tortuosity types) unfolded in parallel over a long period. These physics-based branches are described in separate subsections. The geometric tortuosity types, which appeared more recently, are then presented in the following subsections. The classification of tortuosity is made even more complicated because there exist also mixed tortuosity types. They are mixed in the sense that they fulfil both criteria, for classification as physics-based, and also for classification as geometric tortuosity.

To establish the basis for more precise descriptions of tortuosity, we thus present a *new classification scheme* that uses the two mentioned criteria (i.e., method of determination and concept of definition) for a meaningful distinction of all existing tortuosity types. This results in no more than *three main categories:* (a) *direct geometric,* (b) *indirect physics-based and* (c) *mixed tortuosity types.* This classification scheme also serves as a basis for a *systematic and specific nomenclature,* which aims to provide all relevant information that is necessary for a scientifically correct treatment of the different tortuosity types.

2.2 Hydraulic Tortuosity

2.2.1 Classical Carman-Kozeny Theory

2.2.1.1 Capillary Tubes Model by Kozeny

For porous media, the volumetric flow (Q) induced by a pressure gradient ($\Delta P/L_0$) can be described by Darcy's law from 1856 [6]

$$Q = v_s A = -\frac{\kappa A}{\mu} \frac{\Delta P}{L_0}, \tag{2.2}$$

with superficial flow velocity (v_s), cross-section area (A), dynamic viscosity (μ) and permeability (κ). All resistive effects of the microstructure are implicitly and indistinguishably contained within the permeability (κ). 'In early times', when 3D methods for microstructure investigation were not yet available, flow and its relationship to the underlying microstructure were modeled based on a simplified geometrical model consisting of a bundle of parallel tubes (i.e., equivalent channel model, see Kozeny [5]).

Capillary flow in a straight tube can be described with the Hagen-Poiseuille equation

$$v_c = -\frac{r^2}{8\mu} \frac{\Delta P}{L_0}, \tag{2.3}$$

with capillary velocity (v_c, also called interstitial velocity) and tube radius (r). The comparison of Eqs. 2.2 and 2.3 reveals that permeability (κ) in tube models scales with $r^2/8$.

For models where capillary tubes are not straight, the effective length of the capillary flow path (L_{eff}) is larger than the direct length (L_0), which leads to a reduction of the effective pressure gradient. To correct this effect, Kozeny introduced the notion of hydraulic tortuosity, which he defined as the ratio of the effective hydraulic path length over the direct length (i.e., $\tau_{hydr} = L_{eff\_hydr}/L_0$). This leads to

$$v_c = -\frac{r^2}{8\mu} \frac{\Delta P}{L_{eff_{hydr}}} = -\frac{r^2}{8\mu\tau_{hydr}} \frac{\Delta P}{L_0}. \tag{2.4}$$

To adapt Poiseuille's description of a single tube for equivalent tubes (as analogy for porous media), it is necessary to also consider the impact of pore volume fraction on superficial velocity and associated volume flow. According to Dupuit's relation, superficial velocity (v_s in Eq. 2.2 for porous media flow) is equal to the capillary velocity (v_c in Eq. 2.4 for tube flow) multiplied by porosity (i.e., $v_s = v_c \, \varepsilon$). In analogy to Darcy's law, the equation for volume flow in the capillary tubes model thus becomes

$$Q = v_s A = -\frac{r^2 \varepsilon A}{8\mu T_{hydr}} \frac{\Delta P}{L_0}.$$ (2.5)

The distinct notation of tortuosity factor (T, instead of τ) in Eq. 2.5 originates from a later extension of Dupuit's relation by Carman [2], which is discussed below in context with Eq. 2.14.

Permeability strongly depends on the effective hydraulic radius (r_{hydr}), which represents a tube equivalent radius that is characteristic for the overall viscous drag. Kozeny introduced the hydraulic radius as the ratio of area open to flow (in a 2D cross-section perpendicular to flow) over the perimeter of this area exposed to flow. For a given volume of porous media, the hydraulic radius can also be defined as the ratio of the pipes volume open to flow over the corresponding surface area of these pipes. For a single straight tube, the hydraulic radius is thus half of the tube radius ($r_{hydr\_tube} = \pi r^2 L / 2\pi r L = r/2$).

In a more generalized description for porous media, the volume-to-surface ratio is rewritten as the ratio of porosity over specific surface area per volume ($r_{hydr\_K} = \varepsilon/S_V = r/2$, with subscript K for Kozeny). For non-circular tubes, Kozeny additionally introduced a shape correction factor (c_K). This leads to the well-known semi-empirical Kozeny equation [5]

$$Q = v_s A = -\frac{r^2_{hydr_K}}{c_K} \frac{\varepsilon}{T_{hydr}} \frac{A}{\mu} \frac{\Delta P}{L_0} = -\frac{\varepsilon^3 A}{c_K S_V^2 T_{hydr} \mu} \frac{\Delta P}{L_0}.$$ (2.6)

For the specific case of circular tubes, the Kozeny factor c_K is equal to 2. For non-circular tube cross-sections, shape correction factors in the range from 1.5 to 2.6 were specified based on experimental data.

Combining Eq. 2.6 with Eq. 2.2, we obtain an expression for permeability in terms of porosity, Kozeny factor, specific surface area (per volume) and hydraulic tortuosity factor. This expression, also called Kozeny equation in the literature, reads as

$$\kappa = \frac{\varepsilon^3}{c_K S_V^2 T_{hydr}}.$$ (2.7)

2.2.1.2 Packed Spheres Model by Carman

In 1937, Carman [2] presented some important modifications of Kozeny's equations, in order to describe permeability in granular materials (instead of a bundle of parallel tubes). For this purpose, Carman considered a simplified geometrical model of packed spheres. Specific surface area per total volume (S_V) is replaced by surface area per solid volume (a_V), which then requires the solid volume fraction ($1 - \varepsilon$) as a correction term. For mono-sized spheres, the surface area per solid volume (a_V) can be written as a function of the particle diameter ($a_V = 6/D_p$). For non-spherical particles, the hydraulic radius needs to be corrected with a shape factor (c_C, with subscript C for Carman). With these corrections for shape (c_C) and solid volume fraction ($1 - \varepsilon$), one obtains

$$r_{hydr\_C} = \frac{c_C \varepsilon}{a_V(1 - \varepsilon)} = \frac{c_C D_p \varepsilon}{6(1 - \varepsilon)}. \tag{2.8}$$

Permeability of a packed spheres model (as analogy for granular materials) is thus described with the Carman-Kozeny equation

$$\kappa = \frac{r_{hydr\_C}^2 \varepsilon^2}{2T_{hydr}} = \frac{c_C^2 \varepsilon^3}{2a_V^2(1 - \varepsilon)^2 T_{hydr}} = \frac{c_C^2 D_p^2 \varepsilon^3}{72(1 - \varepsilon)^2 T_{hydr}}. \tag{2.9}$$

Thereby the Kozeny factor (c_K) for tube shapes becomes obsolete and can be replaced with a constant value of 2. The Carman factor (c_C) for correction of non-spherical particle shapes was determined experimentally for grain-sorted powders, whereby values in the range from 0.28 (for mica) to 1 (for spherical particles) were obtained.

2.2.1.3 Different Concepts of Flow Velocity

Carman [2] pointed out that the comparison of Eq. 2.2 (Darcy, porous media, superficial velocity) with Eq. 2.3 (Hagen-Poiseuille, tube flow, capillary velocity) requires a careful consideration of the involved velocities. In principle, velocity can be defined as the ratio of path length over characteristic residence time (t) during which a particle is travelling from inlet to outlet. Three different velocities must be distinguished in context with porous media flow, which was later also discussed by Epstein in 1989 [3]. From the relationship between the three flow velocities, a new definition of hydraulic tortuosity as well as an extension of Dupuit's relation can be deduced, as illustrated in Fig. 2.1:

(a) *Capillary velocity* (v_c)

The *interstitial microscopic* capillary velocity (v_c) used in Eq. 2.3 for tubes is also called intrinsic velocity for porous media. The notion of capillary velocity is based on a microscopic consideration of particles travelling through porous media (or through a single tube). Their capillary velocity can be described as the ratio of the effective tortuous path length, denoted by L_{eff_hydr}, over the residence time (t), i.e.

$$v_c = \frac{L_{eff\_hydr}}{t}. \tag{2.10}$$

Thereby, L_{eff_hydr} is interpreted as a mean length, which is characteristic for a large number of particle pathways. Subsequently, homogenized 'mean' properties of locally defined quantities are denoted with angle brackets, i.e., $<x>$ denotes the homogenized mean of x. For complex porous media, it was not possible for a rather long time to measure the mean length of hydraulic transport pathways or streamlines ($<L_{eff\_hydr}>$). However, a simpler approach to determine the mean capillary velocity $<v_c>$ without measuring L_{eff_hydr} was later presented by Duda et al. [7] and Matyka and Koza [8]. These authors derived mean velocity $<v_c>$ based on velocity vector fields computed with transport simulations at pore scale. These authors then used $<v_c>$ as a basis for calculating the volume averaged tortuosity in an elegant way (see the discussion in context with Eq. 2.18).

(b) *Axial velocity* (v_x)

The *interstitial axial* velocity (v_x) is based on a *macroscopic* observation of flow in porous media, where only the direct path length (L_0), i.e., the sample length between inlet and outlet planes is known. The residence time (t) is the same as for the microscopic observation related to capillary velocity. Then, v_x is given by the ratio of direct path length over residence time, i.e.

$$v_x = \frac{L_0}{t}. \tag{2.11}$$

Capillary velocity (v_c) and interstitial axial velocity (v_x) are equivalent only for the case of a single, straight tube, where $L_{eff\_hydr} = L_0$. In all other cases capillary velocity is higher than axial velocity (i.e., $v_c, \geq v_x$).

(c) *Superficial velocity* (v_s)

Finally, in a porous media, the *macroscopic* superficial velocity (v_s) in Darcy's law (Eq. 2.2) can be deduced from the ratio of volume flow over cross-section area (i.e., $v_s = Q/A$). According to Dupuit's relation, the macroscopic superficial velocity in porous media (v_s) can also be obtained from the interstitial axial velocity (v_x, e.g., in a single tube or in porous media consisting of tubes) with correction of the volume effect using porosity (ε). In this way one obtains

$$v_s = \frac{Q}{A} = v_x \varepsilon = \frac{L_0}{t} \varepsilon. \tag{2.12}$$

The careful distinction of three different flow velocities leads to two main conclusions:

First, from the ratio of Eq. 2.10 over Eq. 2.11, Carman obtained *two equivalent definitions of hydraulic tortuosity*,—first as *ratio of mean path lengths* and second as *ratio of mean velocities*, i.e.

$$\tau_{hydr} = \frac{\left\langle L_{eff\_hydr} \right\rangle}{L_0} = \frac{\langle v_c \rangle}{\langle v_x \rangle}. \tag{2.13}$$

For a long time, conventional definitions of tortuosity focused on the ratio of path lengths. However, with the rise of numerical simulations the consideration of mean velocity components has gained importance as an equivalent definition for tortuosity (see Eq. 2.18).

Second, by combining Eqs. 2.10–2.13, Carman also obtained

$$v_s = v_c \varepsilon \frac{1}{\tau_{hydr}}, \tag{2.14}$$

as an *extension of Dupuit's relation*. Substituted in Eq. 2.5, tortuosity was thus introduced for a second time in context of flow equations (i.e., *first, for the correction of pressure gradient by Kozeny* and *second, for the correction of velocity by Carman*). Consequently, the meaning of the tortuosity factor (T) in Eqs. 2.5–2.9 must be redefined as hydraulic tortuosity by a power of 2 (see [2, 3, 9]), i.e.,

$$T_{hydr} = \left(\frac{L_{eff\_hydr}}{L_0} \right)^2 = \tau_{hydr}^2. \tag{2.15}$$

2.2.2 From Classical Carman-Kozeny Theory to Modern Characterization of Microstructure Effects

2.2.2.1 Limitations of the Carman-Kozeny Approach

Th Carman-Kozeny equations describe two main transport limitations arising from the microstructure. First, viscous drag induced by wall friction is captured with the hydraulic radius (r_{hydr}). Second, non-viscous effects are attributed to reduced pore volume fraction and/or to increased length of transport pathways. These non-viscous effects are described with dimensionless microstructure descriptors for porosity (ε) and hydraulic tortuosity (τ_{hydr}). Variations of tortuous path lengths affect both, superficial velocity, and effective pressure gradient, and thus, tortuosity appears with a power of 2 in the Carman-Kozeny equations.

The Carman-Kozeny equations were introduced at a time when tomography and 3D image analysis were not yet available and therefore hydraulic radius and tortuosity could not be measured directly from the microstructure. As a loophole to this problem, Carman considered a simplified geometrical model consisting of mono-sized spheres as an analogy for the complex pore structure in granular media. For this simplified model the hydraulic radius can be described with easily accessible geometric descriptors ($\varepsilon, S_V, a_V, D_p$), as summarized in previous sections. However, the determination of hydraulic tortuosity remained a major problem. Based on geometric analysis of streamlines in a packed bed of spheres, Carman proposed to use a constant value of $\sqrt{2}$ for τ_{hydr}.

Experimental validations confirmed that the Carman-Kozeny equations are capable to predict permeability and flow reasonably well for simple granular media consisting of mono-sized spheres. For non-spherical particles, Carman introduced a shape correction factor (c_C), which must be fitted for different particle shapes and size distributions separately. It turns out that the uncertainties of permeability predictions with the Carman-Kozeny equations increase with geometric complexity of the granular material (e.g., for non-spherical particles, for wide particle size distributions and for anisotropic particle packing and grain orientation). In the meanwhile, numerous studies have shown that the semi-empirical Carman-Kozeny approach is highly uncertain for materials with complex microstructures (see e.g., [10–13]). Despite these uncertainties, the Carman-Kozeny equations are still widely used for the study of granular materials such as battery electrodes (materials science) and sandstones (geoscience), where they give reasonable predictions of permeability and flow. As will be discussed in Chap. 5, new equations using new morphological descriptors from 3D analysis have been presented in literature, which provide reliable predictions of flow and permeability also for porous media (granular and non-granular) with more complex microstructures.

2.2.2.2 Controversy About (Un)realistic Values for Hydraulic Tortuosity

Much effort was expended to visualize the streamlines of flowing liquids in porous media and to estimate the associated streamline tortuosity. Already in 1956, Carman [14] was able to visualize streamlines by injecting dye into dense packed glass spheres. With this experiment he demonstrated that on average the streamlines diverge from the direction of macroscopic flow by an angle of about 45°. Based on these observations Carman concluded that the *hydraulic streamline tortuosity* ($\tau_{hydr\_streamline}$) in porous granular media must be approximately $\sqrt{2}$. Thus, in early theoretical work, hydraulic tortuosity was often replaced by a constant value of $\sqrt{2}$.

Contrariwise, in experimental work, tortuosity is usually calculated indirectly from relative properties at macroscopic scale. A relative property is defined as ratio of the effective property (e.g., effective electrical conductivity (σ_{eff}) of a porous medium saturated with electrolyte) over the intrinsic property of the transporting medium (e.g., intrinsic conductivity of the pure electrolyte (σ_0)), which results in $\sigma_{rel} = \sigma_{eff}/\sigma_0$. A simple relationship between microstructure and the macroscopic relative property is then often assumed, according to which, for example, the relative conductivity (σ_{rel}) depends only on porosity (ε) and electrical tortuosity (i.e., $\sigma_{rel} = \varepsilon/\tau_{ele}^2$). Hence, when relative conductivity is known from experiment or simulation, the *indirect electrical tortuosity* can then be calculated easily ($\tau_{indir\_ele} = \sqrt{(\varepsilon/\sigma_{rel})}$). By assuming the same simple relationship for relative diffusivity, the *indirect diffusional tortuosity* can be determined in the same way, i.e., by $\tau_{indir\_diff} = \sqrt{(\varepsilon/D_{rel})}$.

For flow and permeability, the micro–macro relationship is more complex since it involves additional microstructure descriptors for the viscous effects (i.e., r_{hydr}). For example, the Carman-Kozeny formulations could be used for calculation of indirect hydraulic tortuosity (i.e., by combining and reformulating Eqs. 2.6, 2.7, 2.9, 2.15).

The resulting expression for *indirect hydraulic tortuosity* then reads as follows

$$\tau_{indir\_hydr} = \sqrt{\frac{r_{hydr}^2 \varepsilon}{\kappa}} = \sqrt{\frac{\varepsilon^3}{C_K S_V^2 \kappa}}. \tag{2.16}$$

However, this equation is rarely used because the involved descriptors are more difficult to determine. For simplicity, the indirect 'hydraulic' tortuosity is thus often calculated with the same simple approach as described above for relative conductivity or relative diffusivity. It is important to note, that the computed values that are reported in literature for indirect tortuosities are usually much higher than $\sqrt{2}$, and sometimes even up to 20 [15–18].

In Chap. 3, we present an extensive collection of empirical data from literature, which is the basis for a systematic comparison of different tortuosity types. This collection of literature data illustrates a clear mismatch between the relatively high values ($\gg 2$) for indirect tortuosities (calculated from known effective properties) versus relatively low values in the range of $\sqrt{2}$ for streamline-tortuosities (from simulated 3D flow patterns). The latter fits well with the predictions from Carman [14]. A possible explanation for this mismatch is given below in Sect. 2.2.2.4.

2.2.2.3 New Methods for Characterization of Hydraulic Tortuosity

(a) *Hydraulic streamline tortuosity* ($\tau_{mixed\_hydr\_streamline}$)

Over the last two decades considerable progress was achieved in tomography, 3D image processing and pore scale modeling. This allows for a computation of the 3D geometry of streamlines based on simulated flow fields and the associated effective path lengths can be described statistically (see e.g., [19–23]), as schematically illustrated in Fig. 2.2. However, to extract a physically relevant mean value for the effective path length (L_{eff}), the question arises how the single streamlines must be counted in the statistical analysis? Bear [24] and Clennell [9] argued that *hydraulic streamline tortuosity* should be calculated by weighting the streamlines with the corresponding fluid fluxes, i.e.,

$$L_{eff\_weighted} = \frac{\sum_i L_i w_i}{\sum_i w_i} \tag{2.17}$$

where w_i represents a weighting factor for the flux represented by streamline i. In the meanwhile, several weighting approaches were presented in literature (see e.g., [19–23, 25]). For a detailed discussion we refer to Duda [7], who concluded that these different weighting approaches lead to inconsistent results. In particular, circular Eddy-currents may impose a significant source of error. Finally, statistical analysis of streamline geometry is computationally expensive, which is a further drawback of this type of tortuosity.

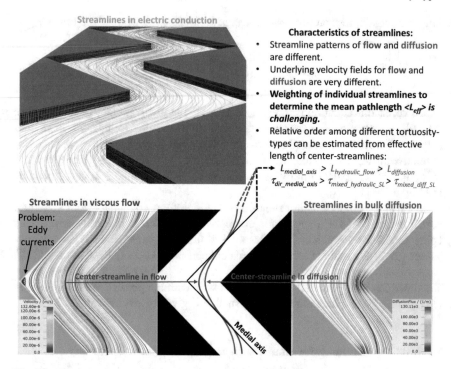

Fig. 2.2 Illustration of streamline tortuosity in a simplified structure (i.e., channel with constant width). For statistical analysis and for determination of a mean effective path length ($<Leff>$), the individual streamlines must be weighted, which is a challenging task. Different streamline patterns are shown for viscous flow (bottom left), for bulk diffusion (bottom right) and for electrical conduction (top left). Colors of the streamlines represent the underlying velocity field. Bottom middle shows the center-streamlines for flow (blue) and diffusion (red) and compares them with the medial axis (black line). From this comparison the following order of tortuosities can be estimated: $\tau_{dir\_medial\_axis} > \tau_{mixed\_hydraulic\_SL} > \tau_{mixed\_diff\_SL} (= \tau_{mixed\_ele\_SL})$

Remark I For more complex microstructures, this order may be different.

Remark II Simulations of electrical conduction with Ohms law and bulk diffusion with Fick's law are mathematically identical. Hence, the electrical and diffusional streamline tortuosities are identical.

(b) *Hydraulic volume averaged tortuosity* ($\tau_{mixed\_hydr\_Vav}$)

A much easier method to compute hydraulic tortuosity was then presented by Matyka and Koza [8] and Duda et al. [7], based on earlier work from Koponen et al. [20]. Instead of focusing on the challenging analysis and weighting of streamlines, their method is based simply on the integration of local vector components from the 3D velocity field:

$$\tau_{mixed\_hydr\_Vav} = \frac{\langle v_c \rangle}{\langle v_x \rangle} = \frac{\int_V v_c(r)d^3r}{\int_V v_x(r)d^3r} \tag{2.18}$$

$$\approx \frac{\sum_{k=1}^n \frac{1}{n}\sqrt{v_x(k)^2 + v_y(k)^2 + v_z(k)^2}}{\sum_{k=1}^n \frac{1}{n}v_x(k)} = \frac{\sum_{k=1}^n \sqrt{v_x(k)^2 + v_y(k)^2 + v_z(k)^2}}{\sum_{k=1}^n v_x(k)},$$

where n is the number of discrete control volumes with equal volume (e.g., voxels from tomography and from the simulated flow field, respectively).

It must be emphasized, that this definition of hydraulic tortuosity is compatible with an alternative definition from Carman (see Eq. 2.13 and Fig. 2.1), who described tortuosity also as the ratio of capillary velocity (v_c) over interstitial axial velocity (v_x). According to Matyka and Koza [8], $<v_c>$ is the 'average magnitude of the intrinsic velocity over the entire pore volume' (i.e., mean capillary velocity) and $<v_x>$ represents the 'volumetric average of the velocity component parallel to the macroscopic flow direction' (i.e., the mean interstitial axial velocity). The mean values are obtained by integration of local properties (i.e., vectors components) at each point r in a discretized (mesh- or voxel-based) velocity field, which is obtained from numerical simulation of flow. The vector components in a flow field are schematically visualized in Fig. 2.3.

Compared to the streamline approach, the volume-averaged approach has several important advantages:

- Neither streamline extraction nor weighting of streamlines are necessary.
- Problems with eddy currents are solved in an elegant way.
- Implementation is relatively easy, and computation is relatively cheap.
- This method not only holds for fluid flow, but also for other transport processes such as diffusion and electrical or thermal conduction.

In literature, this type of tortuosity is called *area (2D) or volume (3D) averaged tortuosity*. For 2D-cases it was introduced by Koponen, 1996 [20]. For 3D-cases it was first applied in 2011 by Matyka and Koza [8], Duda et al. [7] and Ghassemi and Pak [26]. Since then, it is increasingly used for characterization of all kinds of porous media (see e.g., [27–34]).

Throughout the present article, the volume averaged as well as the streamline tortuosities are denoted as *'mixed' tortuosities* (i.e., $\tau_{mixed\_hydr\_Vav}$, $\tau_{mixed\_hydr\_streamline}$). The term 'mixed' emphasizes the fact that this category incorporates 'mixed' information. First, it includes geometric information from 3D analysis of simulated flow fields. Second, it also includes physics-based information from simulation of a specific transport process (i.e., flow, diffusion, or conduction). Thereby, the mixed information is neither determined directly from microstructure nor indirectly from effective or relative properties. (Note: A new tortuosity-classification with direct, indirect, and mixed tortuosities is introduced in Sect. 2.5, see Fig. 2.8).

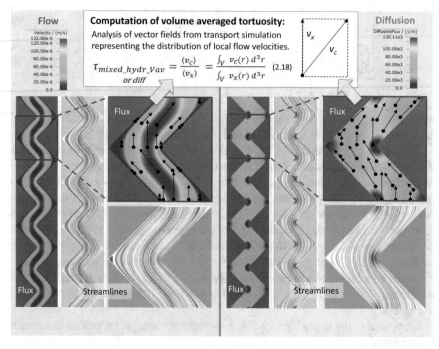

Fig. 2.3 Visualization of principle approach for computation of volume averaged tortuosity ($\tau_{mixed\_hydr\_Vav}$ or $\tau_{mixed\_diff\_Vav}$). Transport simulations of flow (left) and diffusion (right) are performed for a zic-zac channel of constant width. The color code in the plots of streamlines and flux represent the local variation of transport velocities. Note the difference in the flow velocity pattern between flow and diffusion. The volume averaged tortuosity is based on the integration of the velocity vector components in local flow direction (v_c, i.e., capillary velocity) and in direct flow direction (v_x, i.e., axial velocity)

2.2.2.4 New Microstructure Descriptors for Bottleneck Effect and Constrictivity

As mentioned in Sect. 2.2.2.2, the values measured for mixed tortuosities (i.e., volume averaged tortuosity ($_{mixed\_hydr\_Vav}$) and streamline tortuosity ($\tau_{mixed\_hydr\_streamline}$)) are roughly compatible with Carman's estimation of hydraulic tortuosity (ca. $\sqrt{2}$). In contrast, the relatively high values for indirect tortuosities (τ_{indir_hydr} or τ_{indir_ele}) reported in literature indicate that the effective path lengths are overestimated with this approach. By computing tortuosity indirectly from effective transport properties, other limiting effects in addition to path lengths are also included in the calculation of the indirect tortuosity, which explains the obvious overestimation of tortuosity and path lengths. In particular, the limitations arising from narrow bottlenecks are not addressed separately with the indirect approach. The omission of the bottleneck effect is also a major shortcoming of the Carman-Kozeny theory.

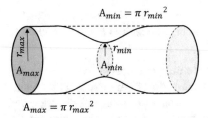

Constrictivity (β) in idealized structures
(i.e. tube flow)

Petersen [35]

$$\beta_{Petersen} = \frac{A_{max}}{A_{min}} = \frac{r_{max}^2}{r_{min}^2} \qquad (2.19)$$

Fig. 2.4 Illustration of a bottleneck in a circular tube. Petersen introduced constrictivity as a correction parameter in transport equations, which accounts for the resistive effects of bottlenecks in pipe flow

(a) *Constrictivity in idealized microstructures by Petersen*

It was shown by Petersen [35] in 1958 that the retarding impact of varying cross-sections for flow in a straight tube can be described with a so-called constrictivity factor ($\beta_{Petersen}$), which he defined as the ratio of cross-section areas at open (A_{max}) and at constricted locations (A_{min}), i.e.,

$$\beta_{Petersen} = \frac{A_{max}}{A_{min}} = \frac{r_{max}^2}{r_{min}^2}, \qquad (2.19)$$

where r_{min} and r_{max} denote the radii of the disk-shaped cross-sections A_{min} and A_{max}, respectively. For the simple case of a constricted pipe, the bottleneck effect and the associated microstructure descriptors (β, r_{min}, r_{max}) are visualized in Fig. 2.4.

(b) *Constrictivity in complex microstructures by Holzer*

In recent years, it became more and more accepted that the bottleneck effect (i.e., constrictivity) is an important retarding effect for transport in porous media, which needs to be considered separately from and/or in addition to the path lengths effect (i.e., 'true' tortuosity) (see e.g., [36–43]). However, until recently there were no methods available to quantify constrictivity (β) from complex microstructures. A suitable method was then introduced by Holzer et al. [38] in 2013, which was later formalized in the framework of stochastic geometry [44]. Thereby, the average sizes of bulges and bottlenecks are obtained *from two different size distribution curves*, see Münch and Holzer [45]:

(a) *continuous pore size distribution (c-PSD)*, for which there is a one-to-one relationship with the granulometry function [46], is used to characterize the *size distribution of bulges*, and

(b) a geometrically defined *mercury intrusion pore size distribution (MIP-PSD,*
     also called porosimetry curve) is used to characterize the *size distribution of*
     bottlenecks.

With the MIP-PSD method, the 3D distance map is modeled in such way that
the pore/phase sizes are evaluated in transport direction from inlet to outlet plane.
Thereby, the pore/phase sizes can only become smaller in transport direction and
therefore the smallest bottleneck 'upstream' represents a hard constraint for the
maximum size in the 'downstream' domains. This geometric treatment is comparable
with the pressure loss that occurs upon viscous flow in porous media, which is also
inverse proportional to the involved bottleneck sizes (r_{min}^2, respectively). Further
details on c-PSD and MIP-PSD can be found in Münch and Holzer [45], and for
further reading about constrictivity see [37, 41, 44].

Figure 2.5 (left) illustrates the concept of the two size distribution methods. The
mean radii corresponding to the volumetric 50% quantiles (i.e., r_{50}) of these two pore
size distributions are considered as mean effective sizes for bulges ($r_{50\_cPSD} = r_{max}$)
and bottlenecks ($r_{50\_MIP\_PSD} = r_{min}$), respectively. Constrictivity (β) is then defined
as the ratio of the squared effective bottleneck radius (r_{min}) over the squared effective
bulge radius (r_{max}), which is the inverse of Petersen's definition of constrictivity, i.e.,

$$\beta = \frac{A_{min}}{A_{max}} = \frac{r_{min}^2}{r_{max}^2} = \frac{1}{\beta_{Petersen}}. \tag{2.20}$$

It is important to note that r_{min} and r_{max} do not describe minimum and maximum
radii of a pore structure, but they represent mean values of two different size distribution curves capturing the sizes of either narrow bottlenecks (*MIP-PSD,* r_{min}) or wide
bulges (*c-PSD,* r_{max}). For further reading about constrictivity we refer to [37, 41, 44].

The fact that constrictivity, still today, is often not included in the traditional
transport equations explains the relatively high values that are typically obtained
when calculating indirect tortuosity from effective transport properties, e.g., with
Eq. 2.16 [38, 41]. In Sects. 2.2.2.2 and 2.2.2.3, a striking discrepancy between mixed
tortuosities (with characteristic values in the range of $\sqrt{2}$) and indirect tortuosity
(with values typically > 2) was described. This discrepancy can mainly be attributed
to the exclusion of constrictivity (bottleneck effect) from the calculation of indirect
tortuosity, which is well documented by Wiedenmann et al. [42]. Indirect tortuosities
that are derived from relative or effective properties (e.g., permeability, conductivity)
are thus often also interpreted as fudge factors (or structure factors), because they
represent an overall resistive effect of the microstructure, which is not or only partly
related to the lengths of transport pathways (see e.g., Clennell [9]).

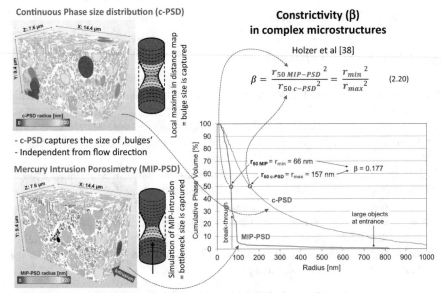

Fig. 2.5 Constrictivity (β) in materials with complex microstructures can be derived based on two different pore size distribution (PSD) methods, which is illustrated for an SOFC electrode consisting of porous LSC (modified after Holzer et al. [38]). To describe the resistive effects from narrow bottlenecks towards electrical transport in the solid phase (i.e., LSC), the same geometric descriptors (i.e., PSD, β) and associated image processing tools can be used as for the mass transport resistance in the pore phase. Two 3D images on the left side represent so-called distance maps with color coded radii of the LSC-phase, from which the c-PSD and MIP-PSD curves on the right side are derived. The continuous phase/pore size distribution curve (c-PSD) captures the size of bulges in the contiguous LSC-phase network (red curve). In analogy, the PSD curve from mercury intrusion porosimetry (MIP-PSD) captures the sizes of bottlenecks in the same contiguous LSC-phase network (blue curve)

In summary, based on improved methods for 3D image analysis over the last two decades, new descriptors for microstructure characteristics have become available. These innovations include new types of tortuosities (direct-geometric and mixed types) and new approaches for measuring characteristic length, hydraulic radius, bottleneck size and constrictivity. Based on these new descriptors also new expressions for the quantitative relationship between microstructure and effective properties (permeability, conductivity, diffusivity) could be formulated. In contrast to the Carman-Kozeny equations, these new expressions have a high prediction power also for materials with complex microstructures. The mathematical description of micro–macro relationships for transport in porous media and how these equations evolved over time is reviewed in Chap. 5.

2.3 Electrical Tortuosity

2.3.1 Indirect Electrical Tortuosity

The concept of electrical tortuosity (τ_{ele}) was developed since ca. 1940 (see Archie [47]) in parallel to the hydraulic tortuosity concept. The electrical tortuosity describes resistive effects of the microstructure, which limit the effective electrical conductivity (σ_{eff}) and, equivalently, increase the effective electrical resistance (R_{eff}) in porous media.

It must be emphasized that electrical conduction in porous media can take place, either, through the pore phase, which is saturated with a liquid electrolyte (whereas the solid phase is insulating). Or, alternatively, the electrical conduction can take place through the solid phase, which is illustrated in Fig. 2.6 for an SOFC-electrode with conductive LSC-phase. In the following section, we consider the case of a porous material saturated with a liquid electrolyte. Ohm's law can be used to describe the electrical flux (J_{ele}) in the saturated porous media, i.e.,

$$J_{ele} = \frac{1}{R_{eff}} \frac{\Delta U}{L_0} = \sigma_{eff} \frac{\Delta U}{L_0}, \tag{2.21}$$

with the potential gradient ($\Delta U/L_0$) as driving force. The resistive formation factor (F_R) was then defined as the ratio of effective electrical resistance of the porous media (R_{eff}) over the intrinsic resistance of the electrolyte (R_0):

$$F_R = \frac{R_{eff}}{R_0} = \frac{\sigma_0}{\sigma_{eff}} = \frac{1}{\sigma_{rel}} = \frac{1}{M}. \tag{2.22}$$

The inverse of the formation factor is the relative conductivity (σ_{rel}), which is also called microstructure (M)-factor. The difference between effective and intrinsic properties is due to the mentioned resistive effects of the underlying microstructure. According to Archie's law [47], which reads as

$$F_R = \frac{1}{\varepsilon^m}, \tag{2.23}$$

the formation factor can also be described as a power law of porosity (ε) with a so-called empirical cementation exponent (m).

Archie's law is widely used in geo- and soil-science. However, it has limited validity because it relates effective transport properties and associated formation factor to a single microstructure characteristic (i.e., porosity) and it ignores all other morphological effects. In an alternative approach by Wyllie and Rose, 1950 [48] the formation factor was described with an additional microstructure characteristic, namely the so-called structural factor (X_{ele}), defined by

$$F_R = \frac{X_{ele}}{\varepsilon} = \frac{\tau_{ele}^2}{\varepsilon}. \tag{2.24}$$

In analogy with Carman's formulation for flow, the structural factor (X_{ele}) was also considered as an equivalent of the (electrical) tortuosity factor ($T_{ele} = \tau_{ele}^2$). This relationship is nowadays more commonly formulated as

$$\sigma_{eff} = \frac{\sigma_0 \varepsilon}{\tau_{ele}^2}. \tag{2.24b}$$

Hence, the electrical tortuosity (τ_{indir_ele}) can be obtained indirectly by plugging experimental results for the formation factor (or relative conductivity) and porosity into Eq. 2.24, which leads to

$$\tau_{indir\_ele} = \sqrt{F_R \varepsilon} = \sqrt{\frac{\varepsilon}{\sigma_{rel}}}. \tag{2.25}$$

This kind of indirect tortuosity (sometimes also called formation tortuosity) is nowadays very prominent because it can be obtained easily from numerical simulations of electric conduction, e.g., with commercial software like GeoDict from Math2Market [49] or with open-source software such as TauFactor from Imperial College London [50]. It must be emphasized that the determination of this indirect electrical tortuosity does not consider any geometric information. It is therefore by no means a measure for the true length of transport pathways. Whenever the tortuosity is afterwards used to determine effective conductivities and/or the resistive effects of microstructure on effective conductivity, respectively, this is a good way to define tortuosity, though.

Katsube et al. [15] performed an extensive investigation on shales that were saturated with electrolyte. The measured values of F_R were in the range from 140 to > 17,000 and porosities were in the range from < 0.01 to 0.1. The corresponding indirect electrical tortuosities took values in the range from 3.4 to 12. Katsube et al. [15] interpreted these values for τ_{indir_ele} as unrealistically high based on geometric considerations. This pessimistic interpretation is in accordance with Carman's 45° argument for the streamlines and associated estimation of $\sqrt{2}$ for streamline tortuosity. Katsube et al. concluded that the empirical results for the indirect electrical tortuosities are unrealistically high because other important microstructure effects (in addition to ε, τ) are not yet included in Eqs. 2.24 and 2.25. Owen [51] and Dullien [52] argued that the influence from narrow bottlenecks needs to be considered as an additional resistive effect. The formation factor was thus redefined by adding constrictivity (β) from Eq. 2.20 (see also [36]), which results in

$$F_R = \frac{\tau_{ele}^2}{\varepsilon \beta}. \tag{2.26}$$

It must be emphasized that the method for direct measurement of constrictivity from porous media, as presented above in Sect. 2.2.2.4, was only introduced in 2013 (Holzer et al. [38]). Because of a lack of suitable methods, constrictivity was thus not considered—until recently—as a separate microstructure characteristic in the calculation of indirect electrical tortuosity. Consequently, unrealistically high values (> 3) are often reported in literature for the indirect electrical tortuosity. These high tortuosity values must be interpreted as mixed information that includes resistive effects not only from tortuous path lengths but also from narrow bottlenecks.

Nevertheless, nowadays it is more and more accepted that the bottleneck effect and constrictivity should be considered separately from the path length effect. Hence, indirect tortuosity is now sometimes also calculated based on a separate treatment of constrictivity (see e.g. He et al. [53]), i.e.,

$$\tau_{indir\_ele\_II} = \sqrt{\frac{\varepsilon \beta}{\sigma_{rel}}}. \tag{2.27}$$

2.3.2   Mixed Electrical Tortuosities

Today, the recent progress in numerical simulation and 3D image processing opens new possibilities for the computation of other (mixed) types of electrical tortuosity. In analogy to the hydraulic tortuosity discussed above in Sect. 2.2.2.3, also the electrical tortuosity can be extracted from simulated 3D fields of electrical flux. This approach provides either streamline ($\tau_{mixed\_ele\_Streamline}$) or volume averaged tortuosities ($\tau_{mixed\_ele\_Vav}$) (see Matyka and Koza [8] and Duda et al. [7]). For the volume averaged tortuosity, the equation can be rewritten as follows:

$$\tau_{mixed\_ele\_Vav} = \frac{\langle v_c \rangle}{\langle v_x \rangle} = \frac{\int_V v_c(r) d^3 r}{\int_V v_x(r) d^3 r} \tag{2.18}$$

Furthermore, it should be noted, that electrical conduction and associated electrical tortuosity are not limited to porous media saturated with electrolyte. As mentioned earlier, the same principles can be used to describe electrical conduction in solid phases of a porous medium or a composite and thereby analyzing the electrical tortuosity of the conducting phase (e.g., electrical conduction in cermet electrodes of solid oxide fuel cells).

2.4 Diffusional Tortuosity

2.4.1 Knudsen Number

This section mainly deals with tortuosity in context with molecular diffusion (also called bulk diffusion), which is the dominant process in systems where chemical interactions between particles (ions, molecules) are negligible. Typically, this is the case for electrolytes with a high level of dilution and/or with an ideal, inert tracer. Also, gas transport at low pressure in macro- and mesoporous media is often dominated by molecular diffusion. Contrariwise, in systems where advection or surface effects (adsorption or dispersion) are important, molecular diffusion may not be the dominant transport process anymore.

The Knudsen number (K_n) is used to distinguish between different diffusion regimes in porous media. K_n is defined by the ratio of mean free path length (λ) over the characteristic length (L_c), i.e.,

$$K_n = \frac{\lambda}{L_c}. \tag{2.28}$$

The mean free path length (λ) depends on pressure, temperature and on the effective cross-sectional area of the gas species. Typically, it is in the range of 30–200 nm. For example, for air at room temperature and ambient pressure λ is 68 nm [54].

The characteristic length (L_c) is an ill-defined property, but it is usually considered as being equivalent to the characteristic pore radius. Hence, for gas transport in nanoporous media with $r_{50} < 10$ nm, the *Knudsen number is much larger than 1*. In this case, we say that we are in the *Knudsen diffusion regime*, which is controlled by *molecule-wall collisions* [55].

For gas transport in *macro-porous media, K_n is < 1*, meaning that we are in the regime of *bulk molecular diffusion*, which is controlled by molecule–molecule collisions. For all liquid electrolytes, λ is very small (nm or smaller) so that diffusion in porous media is usually controlled by bulk diffusion.

For Knudsen numbers *close to 1*, both transport phenomena must be considered. Bosanquet's approximation [55] or the Dusty Gas Model [56] are then often used to model transport in this *mixed regime*.

2.4.2 Bulk Diffusion

2.4.2.1 Indirect Diffusional Tortuosity

For molecular or bulk diffusion $(K_n < 1)$, Fick's first law can be used to describe diffusional flux (J_D), which is driven by a concentration gradient $(\Delta c/L_0)$. The flux in porous media directly scales with diffusivity (D_{eff}), which is an effective property of the system under consideration (see e.g., Satterfield and Sherwood [57]). More precisely,

$$J_D = -D_{eff} \frac{\Delta c}{L_0}. \tag{2.29}$$

The effective diffusivity (D_{eff}) of a porous medium depends on the intrinsic diffusivity (D_0) of pure electrolyte or pure gas, respectively, and on the resistive effects from the microstructure. The transport limitations from obstacles in the microstructure are quantitatively expressed by the relative diffusivity $(D_{rel}$, also called microstructure factor or M-factor (see also Eq. 2.22)), which is a dimensionless characteristic, i.e.,

$$D_{eff} = D_0 M = D_0 D_{rel}. \tag{2.30}$$

Initially, all resistive effects from the underlying microstructure were attributed to the diffusional tortuosity factor $(D_{rel} = 1/T_{diff}$, with $T_{diff} = \tau_{diff}^2)$, see [57]. Still today some authors prefer this definition of tortuosity, which can then be considered as a global transport resistance (e.g., Elwinger et al., [58]). However, it was recognized very early that diffusion depends on different microstructure effects, which are associated with path length variations as well as with pore volume variations. Hence, D_{rel} was then defined in an analogous way as the relative electrical conductivity, including porosity in addition to tortuosity (in analogy to σ_{rel} in Eqs. 2.22 and 2.25, see [3, 48, 59]), i.e.,

$$D_{rel} = \frac{\varepsilon}{\tau_{diff}^2}. \tag{2.31}$$

This leads to the frequently used indirect diffusional tortuosity defined by

$$\tau_{indir\_diff} = \sqrt{\frac{\varepsilon}{D_{rel}}}. \tag{2.32}$$

Note that the mathematical treatment for numerical simulation of bulk diffusion (Fick's law) and electrical conduction (Ohm's law) is identical. In both approaches the Laplace equation is solved. For completion, it is mentioned here that this analogy also applies to Fourier's law of heat conduction $(Q_{thermal} = -\lambda_{eff} (\Delta T/L_0)$; with $Q_{thermal}$ = thermal flux, λ_{eff} = effective heat conduction). It follows that the impact of pore structure on the effective properties of all three processes (bulk diffusion, electric and thermal conduction) must be identical. In fact, it was reported from several experimental studies that the same values are obtained for indirect electrical and diffusional tortuosities when the same porous media was analyzed [60–62].

Furthermore, in a similar way as discussed previously for the indirect electrical tortuosity, also unrealistically high values for indirect diffusional tortuosity were often reported in empirical studies of diffusion. These high values must be attributed again to the fact that the bottleneck effect is not treated as a separate resistive effect. Following van Brakel and Hertjes [36] and in analogy to the electrical tortuosity, this can be improved by introducing constrictivity (β) to the equation for relative diffusion, i.e.,

$$D_{rel} = \frac{\varepsilon \beta}{\tau_{diff}^2}. \tag{2.33}$$

Nevertheless, still today the indirect diffusional tortuosity is often calculated based on Eq. 2.32, without considering constrictivity separately. We conclude that diffusional tortuosity for systems with $Kn < 1$ (bulk diffusion described with Fick's law) is in principle identical with electrical tortuosity (described by Ohmic conduction), and therefore the limiting effects of pore structures are the same (as discussed by Clennell [9]).

The indirect diffusional tortuosity in Eq. 2.32 can be derived in many ways depending on the method by which the relative diffusivity D_{rel} is determined. For example, D_{rel} can be obtained from diffusion experiments (which can be denoted as $\tau_{indir\_diff\_exp}$). Very often, D_{rel} is obtained from simulation of bulk diffusion in a 3D model representing the pore microstructure (which can be specified as $\tau_{indir\_diff\_bulk}$). Alternatively, D_{rel} can be determined with random walk simulation. The random walk simulation is briefly described below for Knudsen diffusion (Sect. 2.3.3, Eq. 2.34), but of course it can be applied in a very similar way also for the computation of D_{rel} in the bulk diffusion regime. By substituting D_{rel} from random walk simulation into Eq. 2.32 we obtain a third type of indirect diffusional tortuosity (i.e., $\tau_{indir\_diff\_Rwalk}$).

2.4.2.2 Mixed Diffusional Tortuosities

(a) *Streamline and volume averaged tortuosities* ($\tau_{mixed\_diff\_Streamline}$, $\tau_{mixed\_diff\_Vav}$)

In analogy to electrical conduction, more sophisticated tortuosity types can be determined nowadays based on numerical simulation of diffusional flux and 3D image processing. The resulting streamline- and volume averaged tortuosities ($\tau_{mixed\_diff\_Streamline}$, $\tau_{mixed\_diff\_Vav}$) represent more rigorous measures of the diffusive path lengths, since they do not mix with a hidden bottleneck effect, as it is usually the case for indirect tortuosity (τ_{indir_diff}). For the volume averaged diffusional tortuosities (see Matyka and Koza [8] and Duda et al. [7]), the mean capillary velocity $<v_c>$ and the mean axial $<v_x>$ velocity can be computed from simulated vector fields. The corresponding equation can be rewritten in analogy to the volume averaged electrical or hydraulic tortuosity:

$$\tau_{mixed\_diff\_Vav} = \frac{\langle v_c \rangle}{\langle v_x \rangle} = \frac{\int_V v_c(r)d^3r}{\int_V v_x(r)d^3r}. \tag{2.18}$$

(b) *Random walk tortuosity* ($\tau_{mixed\_diff\_Rwalk}$)

In context with diffusional mass transport, random walk methods can be used to simulate Brownian motion of particles. From the random walk simulation, a statistical measure for the displacement of moving particles (i.e., the mean square displacement, MSD) can be extracted. The MSD is proportional to the product of time and intrinsic diffusivity (MSD = f (D_0 t)). In porous media, the particle diffusion is hindered by the obstacles of the pore wall. This limiting effect is quantitatively captured by the random walk tortuosity ($\tau_{mixed\_diff\_Rwalk}$), which is defined as the ratio of MSD in free space over MSD in the porous medium.

In principle, the movements in each direction sum together and therefore, the MSD can be decomposed into the Axial Square Displacements (ASD). The axial tortuosities in x-, y-, and z-directions ($\tau_{x/y/z\_mixed\_diff\_Rwalk}$) can then be calculated from the corresponding ASDs. Pytrax is a simple and efficient random walk implementation for calculating the directional tortuosity from 2D and 3D images (see Tranter et al. [63–68]).

2.4.3 Knudsen Diffusion

In nanoporous materials with $K_n \gg 1$ (i.e., in the Knudsen regime), gas diffusion is controlled by collisions with the pore walls. Numerically, this process can be simulated with the random walk method (see e.g., Babovsky [69]). For each particle the corresponding diffusivity can be calculated from displacement length and travel time. For simulations that are based on a large number of particles and sufficiently long travelling time these calculations result in a homogenized effective Knudsen diffusivity (D_{eff_Kn}). The relative Knudsen diffusivity (D_{rel_Kn}) is then again defined as the ratio of effective over intrinsic Knudsen diffusivity, i.e.,

$$D_{rel\_Kn} = \frac{D_{eff\_Kn}}{D_{0\_Kn}} = \frac{\varepsilon}{\tau_{indir\_Kn}^2}. \tag{2.34}$$

Similar as the relative diffusivity in the bulk diffusion regime, also the relative Knudsen diffusivity is a dimensionless property, which describes the resistive impact of pore microstructure against transport. The corresponding indirect Knudsen tortuosity (τ_{indir_Kn}) is obtained by

$$\tau_{indir\_Kn} = \sqrt{\frac{\varepsilon}{D_{rel\_Kn}}}. \tag{2.35}$$

Note that such computed values for relative Knudsen diffusivity (D_{rel_Kn}) and for the indirect Knudsen tortuosity (τ_{indir_Kn}) strongly depend on the intrinsic Knudsen diffusivity (D_{0_Kn}, sometimes also called apparent diffusivity (D_a)). D_{0_Kn} itself can be computed using characteristic length (L_c) and thermal velocity ($v_{th} = k_b\ T/m$), i.e.,

$$D_{0\_Kn} = \frac{1}{3} L_c v_{th}. \tag{2.36}$$

The characteristic length (L_c) is a rather ill-defined property, which is somehow related to pore size distribution and to the average pore size, respectively. The uncertainty associated with L_c propagates into the relative Knudsen diffusivity and into the associated Knudsen tortuosity (τ_{indir_Kn}), which is critically discussed by Zalc et al. [70].

Unfortunately, to the best of our knowledge, there is currently no method available for a more direct analysis of Knudsen tortuosity based on effective path lengths (L_{eff}/L_0), as it is the case for bulk molecular diffusion, for electric conduction and also for viscous flow (see the discussion of mixed-streamline and -volume averaged tortuosities). Knudsen tortuosity is thus generally determined indirectly and therefore it is difficult to understand Knudsen tortuosity as a resistance that is related to distinct morphological features of the pore structure and to the corresponding length of transport pathways.

In literature, the interpretation of Knudsen tortuosity in context with gas diffusion in nanoporous media is highly controversial. For example, Ferguson et al. [71] used different methods (Random walk, FVM) for modeling transport at continuum scale and in the Knudsen regime, which enabled to compute both tortuosities (i.e., for bulk diffusion and for Knudsen diffusion). Also, Gao et al. [72] compared different modeling approaches (Knudsen, Dusty Gas and Oscillator models) that were used to characterize diffusivity and tortuosity in nanoporous media. Gao et al. emphasizes that the resulting tortuosity is highly dependent on the definition of the characteristic pore size (i.e., characteristic length) and on other experimental parameters (chemical species, temperature, pressure). Therefore, Gao et al. [72] suggested using the coordination number (i.e., average number of vertices that are connected to the nodes) instead of the Knudsen tortuosity to describe the impact of nanoscale microstructure on diffusive transport. This approach leads to more stable results and the physical and geometric interpretations of the coordination number are clearer than the indirect Knudsen tortuosity.

2.4.4 Limitations to the Concept of Diffusional Tortuosity

In nanoporous materials, transport mechanisms very often consist of a superposition of several processes such as bulk molecular diffusion, Knudsen diffusion, viscous flow, adsorption, and surface diffusion (see the examples in [53, 72]). For such cases with a mixed transport mechanism, it seems no longer possible to maintain the initial tortuosity concept as proposed by Kozeny and Carman, which is based either on the ratio of path lengths (L_{eff}/L_0) or on the ratio of velocity components ($<v_c>/<v_x>$), see Eqs. 2.13 and 2.18. It is obvious that the indirect tortuosity, which is then usually applied also for mixed transport mechanisms, has limited value in explaining distinct (geometric) microstructure effects. Also, in this case the indirect tortuosity must be rather understood as a fudge factor that describes the bulk resistive effects from microstructure and should not be interpreted as a measure for effective path lengths.

In very fine-grained, nanoporous media the molecular radii of gas species and the thickness of the surface adsorption layer can be in a similar range as the pore radii. In this case, variations of molecular radii and thickness of adsorption layers can become equally important for effective transport as the pore size and pore structure. Tortuosity in such systems is nowadays often determined with dedicated methods of numerical modeling (e.g., molecular dynamics [53]) or experimental characterization (e.g., NMR [58]), but the link with the initial geometrical tortuosity concept (i.e., with path lengths) is often not clear, since other physical effects (e.g., adsorption) may become dominant for diffusion.

2.5   Direct Geometric Tortuosity

The tortuosity types discussed in previous sections cannot be considered as strictly geometric characteristics of the microstructure. The indirect (hydraulic, diffusional, electrical, thermal) tortuosities are derived from effective properties. They are thus rather interpreted as fudge factors or as parameters that describe the bulk resistive effects of the microstructure. The mixed (streamline, volume-averaged, random walk) tortuosities are derived from the vector fields resulting from transport simulations. The mixed tortuosities for the same 3D microstructure thus vary with the simulated transport mechanism (i.e., diffusion, flow, conduction etc.). Therefore, also the mixed tortuosity types do not represent a direct geometric description of the microstructure itself. In contrast, geometric tortuosity includes a whole group of tortuosity types, which entirely depend on the pore morphology, and which are therefore determined directly from 3D images representing the pore microstructure.

The steadily increasing importance of geometric tortuosity types is triggered by the recent progress in the fields of tomography and 3D image analysis, which is summarized in Chap. 4. Many different approaches to measure geometric path lengths and associated geometric tortuosities can be found in literature. The following section describes the most prominent examples.

It must be emphasized that geometric tortuosities do not consider any information regarding the transport mechanisms. This is in contrast, for example, to streamline and volume averaged tortuosities, which are based on simulations of conduction, diffusion, or flow. This is also in contrast to indirect tortuosities, which are extracted from effective properties that are related to specific transport mechanisms.

2.5.1 Skeleton and Medial Axis Tortuosity

The medial axis tortuosity ($\tau_{dir\_medial\_axis}$) can be considered as a prototype for the family of geometric tortuosities. Therefore, in literature it is often just called 'geometric tortuosity' without further specification (see e.g., Stenzel et al. [41]). In our nomenclature, we allocate it to the group of 'direct' tortuosities since it is derived directly from the 3D microstructure by image analysis, in contrast to the indirect tortuosity types. The computation of medial axis tortuosity is based on several rather complex image-processing steps, which are illustrated in Fig. 2.6: The raw data from tomography is first segmented into its constituent phases. The example in Fig. 2.6a shows a fuel cell anode with the phases nickel, Gd-doped ceria, and pores [73]. For each phase a medial axis skeleton (MAS, see Chap. 5 of Soille [74] for details) is then produced (Fig. 2.6b, c). The shortest pathways through the MAS network, which connect couples of inlet and outlet points, are found e.g., with the help of the Dijkstra algorithm. For propagation algorithms on graphs to determine shortest path lengths, the reader is referred to Jeulin et al. [75]. The voxel-based skeleton is then transformed into a 3D graph, i.e., into a network representation consisting of vertices (nodes) and edges (branches) between them. A small portion of a 3D graph is shown in Fig. 2.6d. Note that further information can be attached to edges and vertices in form of additional characteristics of the local microstructure (e.g., local pore size and bottleneck size of each segment, coordinates, and coordination nr of each node etc.). The analysis of the 3D graph is computationally cheap. It reveals a distribution of medial axis tortuosities for each space direction (Fig. 2.6e). Based on graph analysis, tortuosity information of each pathway can be combined with local pore characteristics, such as the paths orientation. The example from Keller et al. [76, 77] in Fig. 2.6f shows a stereographic projection of pore path orientations in an anisotropic clay rock. The path orientation information is combined with the medial axis tortuosities of each path. Tortuosity values are indicated with a color code. In this example, the pathways parallel to the bedding plane (yz-direction) have lower medial axis tortuosities (2.7, dark blue) compared to the pathways perpendicular to the bedding plane (x-direction: up to 13, yellow), which of course has a strong impact on the anisotropy of the macroscopic permeability.

Reproducibility of medial axis tortuosity among different research groups may be a challenge, since there exist many different methods for skeleton extraction from 3D voxel data (see Soille [74] for a detailed description of different skeletonization algorithms). Throughout the present paper, we use the following nomenclature: If the skeleton is not a medial axis representation, then we speak of $\tau_{dir\_skeleton}$ instead of $\tau_{dir\_medial\_axis}$. The procedure for generation of a medial axis skeleton (MAS) can be found e.g., in Lindquist et al. [78]. MAS generation is typically based on an iterative erosion process called topological thinning. The resulting skeleton consists of lines or curves with a thickness of one voxel. These curves are always located in the center of the pore bodies. As previously mentioned, the shortest pathways between couples of inlet and outlet points can then be algorithmically computed, e.g., by means of

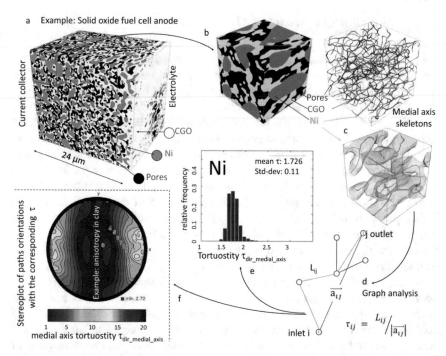

Fig. 2.6 Illustration of image processing steps, which are the basis for computation of medial axis tortuosity ($\tau_{dir\_medial\_axis}$): **a** FIB-SEM tomography and phase segmentation of a cermet anode for solid oxide fuel cells (SOFC) with pores (black), nickel (gray) and Gd-doped ceria (CGO, white) from Holzer et al. [73], **b** zoom-in of **a**, **c** medial axis skeletons (MAS) of each phase, **d** schematic illustration of a graph consisting of branches/edges and nodes/vertices (e.g. used for analyzing path lengths), **e** graph analysis, i.e. statistical analysis of paths lengths and associated medial axis tortuosities (mean τ) for the nickel phase in the main transport-direction (z) and **f** stereoplot showing anisotropic distribution of tortuosities in a clay rock (d + f from Keller et al. [76, 77])

the Dijkstra algorithm [79]. It turns out that for complex pore structures a robust skeletonization procedure is challenging.

Efforts have been made to develop algorithms for an accelerated computation of shortest pathways in skeleton networks. Besides the propagation algorithm in Jeulin et al. [75] mentioned above, TESAR (Tree-structure Extraction algorithm delivering Skeletons that are Accurate and Robust) was introduced by Sato et al. [80]. Thereby the distance field used in the Dijkstra algorithm is modified, such that the shortest pathways in a pore network can be found in a fast and reliable way. This algorithm thus also reveals a medial axis skeleton (MAS). It was later implemented in commercial software, such as Avizo Fire (thermofischer.com), which was used by many authors as a basis for graph analysis to compute medial axis tortuosity (see e.g., [41, 73, 76, 77, 81, 82]; examples are shown in Fig. 2.6). Keller et al. [76] also presented a detailed description of the transformation steps from a voxel-based skeleton, via a 3D graph to a vector-based list of segments and nodes (assigned with local pore characteristics, such as effective lengths, orientation, pore size and bottleneck size).

Open-source software solutions for skeletonization and extraction of medial axes are available e.g., as Matlab code (Tort3D described by Al-Raoush and Madhoun [83]) or in ImageJ/Fiji (see https://imagej.net/Skeletonize3D and imagej.net/Analyz eSkeleton).

Furthermore, Thiedemann et al. [84] also presented methodological details that can be used for an extension of medial axis tortuosity towards weighting of pathways for their fluxes. This approach is based on a detailed analysis of the 3D graph including various local structural characteristics (see also Jungnickel [85] for a general reference to graph theory). Thereby the bottleneck size, which limits the flux of a specific pathway, is used for weighting the segments and local path lengths, which in turn affects the overall statistics of effective path lengths and the associated medial axis tortuosity. The weighting with bottleneck size can be understood intuitively as an analogue of the previously discussed weighting of streamlines by flux. A similar approach was recently also described by Nemati et al. [25].

In summary, the computation of medial axis tortuosity is based on various complex image-processing steps (e.g., skeleton extraction, 3D graph analysis). Therefore, its implementation is complex, and its computation may be time consuming. A major disadvantage is the fact that skeleton extraction can be done in many ways, which may lead to different results for the same type of geometric tortuosity. Moreover, it becomes difficult to interpret the extracted skeletons as transport networks in case of high porosities. Nevertheless, based on fundamental graph theory, this approach allows combining tortuosity with other local characteristics (bottleneck size, pore path orientation, connectivity), which are important for understanding the influence of microstructure on effective transport properties.

2.5.2 Path Tracking Method (PTM) Tortuosity

The path tracking method (PTM) was introduced by Sobieski et al. [86–88]. It allows for a fast computation of a geometric tortuosity (i.e., τ_{dir_PTM}) that is very similar to the skeleton approach discussed in the previous section. However, the PTM algorithm is only applicable for microstructures consisting of packed spheres. It identifies tetragonal structures formed by four neighboring spheres. These tetragons represent an approximation of the interstitial pore space. The algorithm then finds the shortest pathways through the material by connecting gravity centers in the base triangles of neighboring tetragons. The resulting pathways from PTM are similar to the shortest pathways in a network skeleton. Despite this similarity, it must be emphasized that the pathways used for PTM differ in general from the medial axis representation.

In [87], a comparison of the PTM tortuosity (τ_{dir_PTM}) with the hydraulic volume averaged tortuosity ($\tau_{mixed\_hydr\_Vav}$) is presented, which shows almost identical results. However, whereas the geometric PTM approach is fast and easy, the hydraulic volume averaged tortuosity (based on e.g., LBM simulations) is computationally expensive and time consuming. Hence, PTM is a fast method for measuring geometric tortuosity in 3D models of packed spheres or particles.

2.5.3   Geodesic Tortuosity

The concept of geodesic tortuosity (τ_{dir_geod}) relies on the geodesic metric in image data introduced by Lantuéjoul [89] and is used, e.g., in [41, 90, 91] to compute the corresponding geodesic tortuosity. The geodesic tortuosity is based on a statistical analysis of shortest path lengths (L_{eff}) from inlet- to outlet-planes divided by the sample length (L_0). For this purpose, the shortest path lengths are defined in terms of geodesic distances within the set of those voxels that represent the transporting phase (see e.g., Stenzel et al. [41]). Figure 2.7 illustrates the difference between geodesic (red) and medial axis (green) tortuosity. For the medial axis tortuosity there is only one starting point per pore body in the inlet plane, whereas for the geodesic tortuosity all voxels of the transporting phase in the inlet plane are taken as starting points. With increasing distance, the numerous pathways starting from different seed points concentrate on a few geodesic tracks. In both cases (i.e., geodesic and medial axis tortuosities) the shortest pathways can be calculated using the Dijkstra algorithm [79, 92]. However, whereas in the case of medial axis tortuosity the shortest pathways are restricted to the centerlines, for the case of geodesic tortuosity all voxels of the transporting phase are interpreted as vertices of a graph, which are connected to their neighboring voxels (26-neighborhood). Thus, in average, the geodesic pathways are shorter than the medial axis pathways. Note that—due to discretization errors on the voxel grid—single geodesic pathways might be longer than the corresponding paths along the skeleton. In [41] the following relationship

$$\tau_{dir\_geodesic} = 0.76\tau_{dir\_median\_axis} \qquad (2.37)$$

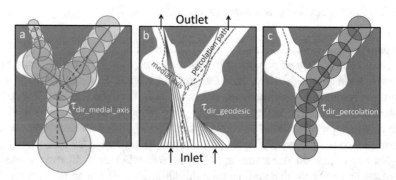

Fig. 2.7 Schematic illustration of geometric pore pathways used to measure path lengths for **a** medial axis tortuosity, **b** geodesic tortuosity and **c** percolation path tortuosity. The percolation method finds the pathways with the least constriction (i.e., pathway, along which the largest possible sphere can migrate from inlet- to outlet-planes). In contrast, the medial axis and geodesic methods find the shortest pathways for each couple of inlet- and outlet-points, independent from the corresponding bottleneck dimensions. Note: Pathways from medial axis and percolation methods are indicated as dotted lines with green and blue colors, respectively, as guides to the eye

has been empirically derived by linear regression, using 43 virtual microstructures generated by a specific type of a 3D stochastic microstructure model. The coefficient of determination R^2 quantifying the goodness of fit for linear regression was equal to 0.81. Note that R^2 is between 0 and 1, where 1 indicates a perfect fit.

A mathematical formalization of geodesic tortuosity in the framework of random sets, a key object in stochastic geometry and mathematical morphology for microstructure characterization [46, 93], was recently provided by Neumann et al. [44, 94], while a slightly modified version of geodesic tortuosity was presented by Barman et al. [11].

2.5.4 Fast Marching Method (FMM) Tortuosity

The fast-marching method (FMM) tortuosity is very similar to the geodesic tortuosity, in the sense that it also considers geodesic distances within the voxel space of a given phase. The FMM algorithm is based on the simulation of a propagating front from inlet- to outlet-plane, which is described e.g., by Vicente et al. [95]. In particular, FMM solves the following Eikonal equation (see Sethian et al. [96–99])

$$\|\nabla T(x)\| F(x) = 1, (F(x) > 0) \qquad (2.38)$$

on the voxel grid, where $T(x)$ is the arrival time at location x and $F(x)$ is the speed of the front. For each voxel the minimum arrival time is computed and by considering the speed of the front, this results in a distance map representing the shortest path lengths (L_{FMM}). For each pixel in the outlet plane, the corresponding FMM tortuosity can be calculated by dividing L_{FMM} through the sample length (L_0).

Jørgensen et al., 2011 [100, 101] describes the FMM method in context with microstructure analysis of SOFC electrodes. Thereby, FMM is also used to extract additional (local) information of the transporting phase network, such as the distribution of interface distances, distribution of characteristic path diameters and identification of dead-end pores. A further application of FMM was presented by Taiwo et al. [102] for battery electrodes. A recipe for the implementation of FMM can be found in Appendix E (supplementary info) of Hamann et al. [103].

In summary, the FMM tortuosity is very similar to the geodesic tortuosity, in the sense that it also finds the shortest (geodesic) pathways within the voxel space representing the transporting phase. In addition, it is computationally cheap and relatively fast.

2.5.5 Percolation Path Tortuosity

Percolation path tortuosity is based on an algorithm that finds the pathway(s) with the least constricting bottleneck(s) (i.e., with the largest minimum bottleneck size).

Hence, this algorithm allows the largest possible sphere(s) to travel from inlet- to outlet-plane and, at the same time, it finds the shortest path through the network for this sphere. Tortuosity is then defined as ratio of percolation path length over direct length ($\tau_{dir\_percolation} = L_{percolation}/L_0$). This method is, for example, implemented in the GeoDict Software (www.math2market.com). Thereby it is possible not only to calculate the percolation path for a single largest sphere but also for a defined number (n) of largest spheres. Hence, it enables us to find the n least constricting pathways and it calculates the corresponding mean tortuosity.

As shown in Fig. 2.7c, the blue sphere (with radius corresponding to the least constricting bottleneck) cannot pass through the narrow bottlenecks of the direct pathway (left) and hence it must take a deviation (right pathway). Percolation tortuosity is thus often larger than medial axis tortuosity, which takes more direct pathways through narrow bottlenecks.

The percolation pathways capture the maximum possible opening, which can be intuitively associated with a pathway of high flux. In contrast, the pathways for medial axis and geodesic tortuosities capture their shortest pathways regardless of the bottleneck radius, and therefore medial axis and geodesic tortuosities do not represent characteristics that can be related to pathways of high flux. Considering percolation path tortuosity for varying radii reveals interesting insights on porous microstructures going beyond the information gained by geodesic tortuosity. This is demonstrated using an example of paper-based materials in [46, 104].

2.5.6 Pore Centroid Tortuosity

For the pore centroid method (see e.g., [81, 101]), the 3D image volume is processed as a stack of 2D images. In each 2D section the position of the center of mass is determined for the transporting phase (e.g., pores). These centers are then tracked in transporting direction perpendicular to the 2D images, which results in one single tortuous centroid path. The centroid tortuosity is then calculated as the ratio of the effective centroid path length (L_{eff}) over the sample thickness (L_0). The pore centroid method is a quick and simple method, which is, e.g., implemented in the Avizo Software (www.thermofischer.com). For increasing volume fractions, the mass center approaches the image center and thus, pore centroid tortuosity goes to one. One can think of simple examples for microstructures with low values of centroid tortuosity (close to one), where the actual transportation paths are highly tortuous. Hence, the relevance of the centroid tortuosity in context with microstructure—property relationships is highly uncertain.

Finally, further approaches for the extraction of geometric tortuosity from 3D images can be found in literature. They are usually based on distance propagation and/or shortest path algorithms (see examples in [105–107]).

2.6 Tortuosity Types: Classification Scheme and Nomenclature

2.6.1 Classification Scheme

The above-presented review reveals a multitude of different tortuosity types. However, in literature, in conference presentations and in associated scientific discussions dealing with tortuosity, the type of tortuosity under consideration is very often not properly defined. This lack of information often becomes the source of confusion and misunderstanding. For clarification, we propose to use a rough classification scheme with only three main categories of tortuosities. For a more precise specification, we introduce a systematic nomenclature that builds on the simple classification scheme. The nomenclature aims to provide all relevant details that are necessary for proper interpretation of the specific tortuosity under consideration. Both, the classification scheme and the detailed nomenclature approach are illustrated in Fig. 2.8.

The classification of tortuosity (Fig. 2.8, top) is based on two main criteria:

(a) *The method of determination:*

The method of determination is in most cases either based on a direct geometric analysis of the microstructure using tomography and 3D image analysis (called direct τ). Alternatively, tortuosity can be deduced indirectly from effective transport properties (called indirect τ).

(b) *The concept of definition:*

The concept of definition is in most cases either based on the assumption that tortuosity and associated path lengths (L_{eff}) are geometric properties of the microstructure (called geometric τ) or, alternatively, the definition emphasizes that tortuosity is a function of the transport process under investigation, such as viscous flow, diffusion, or electric conduction (called physics-based τ).

It turns out that the method of determination and the concept of definition are generally linked with each other in a specific way, which leads to a reduction to three main categories of tortuosity. The first category consists of *direct, geometric tortuosity* types. The second category consists of *indirect, physics-based tortuosity types*. And the third category consists of *mixed types*, including streamline and volume averaged tortuosities (i.e., $\tau_{mixed\_phys\_streamline}$, $\tau_{mixed\_phys\_Vav}$). The definition of mixed types emphasizes both, the dependency on the transport process (i.e., physics-based) as well as the geometric aspect of path lengths and associated tortuosity. Characterization of mixed types is challenging, because the required information cannot be obtained directly from 3D image analysis of the microstructure. First, it requires some numerical 3D simulation to compute a transport-specific volume field of velocities or fluxes. In a second step, 3D image analysis is then used to extract the lengths of streamlines or the volume-averaged ratio of vector components from these volume fields (see Sect. 2.2.2.3).

Tortuosity classification scheme

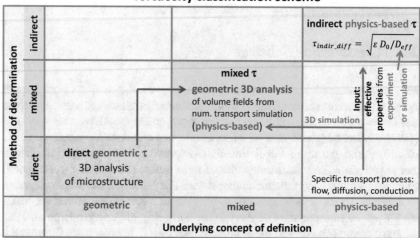

Tortuosity nomenclature

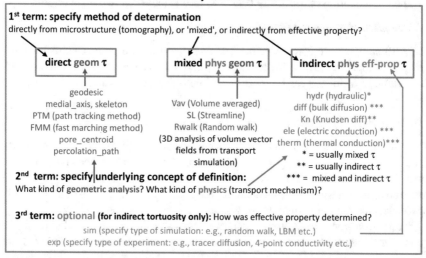

Fig. 2.8 Tortuosity classification scheme (top) and tortuosity nomenclature (bottom). Three main classes of tortuosity can be identified based on the method of determination: direct τ, mixed τ and indirect τ. For precise tortuosity nomenclature, additional information on the underlying concept of definition (geometric, physics-based) and on the effective property characterization method (for indirect τ) is added in the second and third terms. The classification scheme and nomenclature should help avoiding confusion in the discussion and interpretation of tortuosity

2.6.2   Nomenclature

Based on our classification scheme we introduce a new tortuosity nomenclature (Fig. 2.8, bottom), which consists of 2 or 3 terms. With the *first term* we describe the *method of determination*, which is one of three main categories (i.e., direct, mixed, or indirect determination of tortuosity).

The nomenclature then contains additional information on the underlying concept of definition. With the *second (and third) term* we specify details related to the *underlying concept of definition*. This can be either, a specification of the geometric analysis (for direction tortuosities), or a specification of the involved transport process (for indirect tortuosities). For mixed tortuosities, it is necessary to specify both, the geometric analysis, and the involved transport process. In the following section, the nomenclature is rigorously applied to all relevant tortuosity types, which are treated in this book.

2.6.2.1   Category I: Direct Geometric Tortuosities (τ_{dir_geom})

The name includes a *first term (dir)*, which emphasizes direct 3D analysis of the microstructure, e.g., from tomography data. The *second term* (i.e., 'geom' alias medial axis, PTM, percolation, geodesic, FMM or pore centroid) specifies the *geometric concept or method*. The following examples of direct geometric tortuosities were discussed previously in this section:

- $\tau_{dir\_medial\_axis}$ and $\tau_{dir\_skeleton}$
- τ_{dir_PTM} (path tracking method)
- $\tau_{dir\_percolation}$
- $\tau_{dir\_geodesic}$
- τ_{dir_FMM} (fast marching method)
- $\tau_{dir\_pore\_centroid}$.

Note: In a recent review by Fu et al. (2020) [66] different names are used for geometric tortuosities. $\tau_{dir\_geodesic}$ is termed 'direct shortest path searching method' (DSPSM) and $\tau_{dir\_skeleton}$ is termed 'skeleton shortest path searching method' (SSPSM).

2.6.2.2   Category II: Mixed Tortuosities ($\tau_{mixed\_phys\_streamline}$, $\tau_{mixed\_phys\_Vav}$, $\tau_{mixed\_diff\_Rwalk}$)

The *first term (mixed)* defines the category. Mixed tortuosities are neither calculated directly from a morphological analysis of the pore structure, nor are they determined indirectly from effective properties. Typically, mixed tortuosities are obtained by a multi-step process, which starts with a specific simulation of transport, and which is then complemented with an additional postprocessing step (i.e., geometric analysis) of the simulation output. This simulation output can be, for example, a velocity field from simulation of flow, conduction, or diffusion. The output can also consist of many random walkers, which are obtained by simulation of diffusion.

The *second term (physics-based)* thus contains information about the specific transport process under consideration. This can be either viscous flow (hydr), bulk diffusion (diff), electric or thermal conduction (ele, therm).

The *third term* then describes the *geometric method*, which is either based on the analysis of streamlines or volume-averaged vector components (i.e., ratio of velocity vector components) or lengths of random walkers:

- $\tau_{mixed\_hydr\_streamline}$
- $\tau_{mixed\_diff\_streamline}$
- $\tau_{mixed\_ele\_streamline}$
- $\tau_{mixed\_therm\_streamline}$
- $\tau_{mixed\_hydr\_Vav}$
- $\tau_{mixed\_diff\_Vav}$
- $\tau_{mixed\_ele\_Vav}$
- $\tau_{mixed\_therm\_Vav}$
- $\tau_{mixed\_diff\_Rwalk}$.

Note: In literature, tortuosities belonging to categories II (mixed) and III (indirect) are often not distinguished and all of them are called 'flux-based' or 'physical' (see e.g., Tjaden et al. [65] and Fu et al. [66]). Usually the flux-based hydraulic tortuosities are calculated from streamlines or velocity fields (i.e., they belong to the category II: mixed tortuosities), whereas the flux-based electrical and diffusive tortuosities are typically calculated from the corresponding effective properties (i.e., they belong to category III: indirect tortuosity).

2.6.2.3 Category III: Indirect Physics-Based Tortuosities ($\tau_{indir\_phys\_sim}$, $\tau_{indir\_phys\_exp}$)

The *first term (indirect)* defines the category. The *second term (physics-based)* contains information about the *specific transport process* under consideration. The physical nature of transport is either electrical conduction (ele), thermal conduction (therm), bulk diffusion (diff), Knudsen diffusion (Kn) or hydraulic flow (hydr):

- $\tau_{indir\_ele\_sim\ or\_exp}$ $\tau_{indir\_ele} = \sqrt{(\varepsilon/\sigma_{rel})}\ or = \sqrt{(1/\sigma_{rel})}\ or = \sqrt{(\varepsilon\beta/\sigma_{rel})}$
- τ_{indir_therm} $\tau_{indir\_therm} = \sqrt{(\varepsilon/K_{rel})}\ or = \sqrt{(1/K_{rel})}\ or = \sqrt{(\varepsilon\beta/K_{rel})}$
- τ_{indir_diff} $\tau_{indir\_diff} = \sqrt{(\varepsilon/D_{rel})}\ or = \sqrt{(1/D_{rel})}\ or = \sqrt{(\varepsilon\beta/D_{rel})}$
- τ_{indir_Kn} $\tau_{indir\_Kn} = \sqrt{(1/D_{Kn\_rel})}$ (see Eq. 2.35)
- τ_{indir_hydr} $\tau_{indir\_hydr} = \sqrt{(rh^2\,\varepsilon/\kappa)}$ (see Eq. 2.16).

The value of the effective property used as input for indirect tortuosity may be different if it is determined by simulation or with an experimental approach (see e.g., the discussion of apparent tortuosity-discrepancy for Li ion batteries by Usselgio-Viretta et al. [108]). Therefore, we recommend adding a third term that describes the nature of effective property input ('sim' for simulation or 'exp' for experimental).

It is well documented that the indirect tortuosity is dependent on the method by which the underlying effective property is determined. Examples for experimental characterization approaches are diffusion cells, tracer diffusion experiments, measurements of electrical resistance and formation factor, electrochemical experiments (EIS), flow-cell experiments for gases or liquids. It is clear that the measured transport properties may strongly depend on experimental parameters, which then also affects the indirect tortuosity. In a similar way, methodological details of transport simulation will influence the resulting effective properties and associated indirect tortuosity. These aspects cannot be captured by nomenclature and should therefore be described separately.

Furthermore, values obtained for indirect tortuosity are also heavily dependent on the underlying mathematical expression, which describes the relationship between the effective property (input) and tortuosity (output). In most cases, this relation is of the same type as Eqs. 2.25 and 2.32 (e.g., $\tau_{indir\_ele} = \sqrt{(\varepsilon/\sigma_{rel})}$). However, different relationships can be found in literature, either without consideration of porosity ($\tau_{indir\_ele\_inclPore} = \sqrt{(1/\sigma_{rel})}$), where inclPore means that the pore volume effect is included in this indirect tortuosity calculation) or with separate consideration of constrictivity in addition to porosity (see Eq. 2.27: $\tau_{indir\_ele\_exBN} = \sqrt{(\varepsilon\,\beta/\sigma_{rel})}$, where exBN means that the bottleneck effect is excluded from this indirect tortuosity calculation). Hence, for electrical conduction, details of the indirect tortuosity calculation can be expressed as follows:

- $\tau_{indir\_ele} = \sqrt{(\varepsilon/\sigma_{rel})}$ (standard case: Eq. 2.25)
- $\tau_{indir\_ele\_inclPore} = \sqrt{(1/\sigma_{rel})}$
- $\tau_{indir\_ele\_exBN} = \sqrt{(\varepsilon\,\beta/\sigma_{rel})}$.

The cases for thermal conduction and diffusion can be treated analogously. In any case, it is recommended that if the indirect tortuosity for diffusion and conduction is not derived with the standard expression (Eqs. 2.25 and 2.32) this should be emphasized explicitly, and the corresponding mathematical expression should be declared. Finally, it must be emphasized that the indirect tortuosity for flow is rarely used, because it requires knowledge of the hydraulic radius (see Eq. 2.16).

Hence, the *indirect tortuosity* can be determined in many ways, which makes comparisons difficult. To avoid confusion, 'indirect tortuosity' always calls for some *detailed specifications* (in addition to nomenclature) regarding *a) the underlying method used to determine the effective property* by experiment or simulation, and *b) the underlying equation, which relates indirect tortuosity with effective property*, and which is used to calculate the indirect tortuosity. Without these specifications, 'indirect tortuosity' is an ill-defined characteristic.

2.7 Summary

Tortuosity is generally defined as ratio of the mean effective path length over the direct path length ($\tau = <L_{eff}>/<L_0>$). However, determination of the mean effective path length in complex disordered microstructures is not an easy task. Therefore, many different concepts, definitions and methods of characterization can be found in literature. This diversity often leads to confusion and misinterpretation. More than 20 different tortuosity types are presented and discussed in the present review.

In order to enable a precise description of a specific tortuosity, a new and systematic *tortuosity nomenclature* is presented in this chapter. Figure 2.8 can be used as a guide that helps to find the correct name of a specific tortuosity type. This nomenclature is based on a *classification scheme* that uses *two distinctive criteria*:

(a) *The method of determination* distinguishes tortuosities that are either calculated *indirectly* from effective transport properties or *directly* from 3D images representing the microstructure under investigation.

(b) *The concept of definition* makes a distinction between physics-based tortuosities (i.e., hydraulic, electric, diffusional, thermal τ) or geometric tortuosities. The latter represent intrinsic properties of the microstructures, and they are thus independent from the involved transport process.

It turns out that all relevant tortuosity types can be grouped into *three main categories*:

I: Indirect physics-based tortuosities (τ_{indir_phys})
II: Direct geometric tortuosities (τ_{dir_geom})
II: Mixed tortuosities (τ_{mixed_phys}).

I: The indirect physics-based tortuosities describe bulk resistive effects from the microstructure towards a specific transport process (i.e., viscous flow, electric and thermal conduction, bulk and Knudsen diffusion). The indirect physics-based tortuosities typically overestimate the true path length because they are calculated from the corresponding effective properties, which also include other limiting effects from the microstructure (e.g., bottlenecks).

When using the Carman-Kozeny equation for prediction of hydraulic flow in porous media, very often the indirect hydraulic tortuosity is fixed at a value of $\sqrt{2}$ (based on geometric arguments from Kozeny). This approach may work well for granular materials consisting of packed spheres, but it typically fails to predict the transport properties of more complex microstructures. For the latter cases, a more sophisticated consideration of tortuosity is required.

II: The direct geometric tortuosities include a group of metrics that are capable to provide realistic estimations of the mean path length (i.e., τ_{dir_geom}, with 'geom' = medial axis, skeleton, geodesic, fast marching method FMM, path tracking method PTM, percolation path or pore centroid). It must be emphasized that the direct geometric tortuosities consider the path length as an *intrinsic property of the microstructure*, which is *independent from the transport process*.

There is also another important aspect to be considered in context with the direct geometric tortuosities. In order to capture the entire resistive effect from the microstructure, one has to consider also other morphological effects such as the bottleneck effect (constrictivity) and viscous drag at the pore wall (hydraulic radius), in addition to the path length effect (geometric tortuosity).

III: The mixed tortuosities represent the most advanced descriptions of the path length effect. Typically, the mixed tortuosities are based on a numerical simulation of the involved transport process by using as an input the discretized 3D microstructure (e.g., from tomography or from stochastic geometry modelling). Geometric analysis can then be performed on the simulated flow fields. This approach reveals mixed tortuosities that are calculated either from volume averaged velocity vectors, or from mean path lengths of streamlines and/or random walkers.

The volume averaged tortuosity ($\tau_{mixed\_phys\_Vav}$) is perceived as a particularly promising type of tortuosity. It is based on the integration of local velocity vectors over the simulated flow field. Thereby, two specific vector components are considered: one parallel to the *local, microscopic flow direction* (v_c) and one parallel to the *direct, macroscopic flow direction* (v_x). In this way, an alternative definition of tortuosity is obtained, which is *the ratio of the mean capillary velocity over the mean axial velocity* ($\tau_{mixed\_phys\_Vav} = <v_c>/<v_0>$). This approach is computationally cheap and geometrically simple. It provides a reliable description of the mean effective path length also for complex microstructures, and it captures the specific impact of the involved transport process.

References

1. R.B. Bird, W.E. Steward, E.N. Lightfood, *Transport Phenomena*, 2nd edn. (John Wiley & Sons, New York, 2007)
2. P.C. Carman, Fluid flow through granular beds. Chem. Eng. Res. Des. **75**, S32 (1997)
3. N. Epstein, On tortuosity and the tortuosity factor in flow and diffusion through porous media. Chem. Eng. Sci. **44**, 777 (1989)
4. Y.L. Golin, V.E. Karyarin, B.S. Postelov, V.I. Sevedkin, Tortuosity estimates in porous media. Soviet Electrochem. **28**, 87 (1992)
5. J. Kozeny, Über Kapillare Leitung Des Wassers Im Boden. Sitzungsbericht Der Akademie Der Wissenschaften Wien **136**, 271 (1927)
6. H. Darcy, Les Fontaines Publiques de La Ville de Dijon (1856)
7. A. Duda, Z. Koza, M. Matyka, Hydraulic tortuosity in arbitrary porous media flow. Phys. Rev. E Stat. Nonlin. Soft Matter Phys. **84** (2011)
8. M. Matyka, Z. Koza, How to calculate tortuosity easily? AIP Conf. Proc. **1453**, 17 (2012)
9. M.B. Clennell, Tortuosity: a guide through the maze, in *Developments in Petrophysics*, ed. by M.A. Lovell, P.K. Harvey (Geol. Soc. Spec. Publ. No. 122, 1997), pp. 299–344
10. A. Hunt, R. Ewing, B. Ghanbarian, *Percolation Theory for Flow in Porous Media*, vol. 880, 2nd edn. (Springer International Publishing, Cham, 2014)
11. Y. Ichikawa, A.P.S. Selvadurai, *Transport Phenomena in Porous Media* (Springer, Berlin Heidelberg, 2012)
12. D.A. Nield, A. Bejan, *Convection in Porous Media* (Springer, New York, NY, 2013)
13. S.M.R. Niya, A.P.S. Selvadurai, A statistical correlation between permeability, porosity, tortuosity and conductance. Transp. Porous Media **121**, 741 (2018)

14. P.C. Carman, *Flow of Gases through Porous Media* (Butterworth, London, 1956)
15. T.J. Katsube, B.S. Mudford, M.E. Best, Petrophysical characteristics of shales from the Scotian shelf. Geophysics **56**, 1681 (1991)
16. S. Bhatia, Directional autocorrelation and the diffusional tortuosity of capillary porous media. J. Catal. **93**, 192 (1985)
17. C.N. Satterfield, P.J. Cadle, Gaseous diffusion and flow in commercial catalysts at pressure levels above atmospheric. Ind. Eng. Chem. Fundam. **7**, 202 (1968)
18. C.N. Satterfield, P.J. Cadle, Diffusion in commercially manufactured pelleted catalysts. Ind. Eng. Chem. Process. Des. Dev. **7**, 256 (1968)
19. A. Koponen, M. Kataja, J. Timonen, Permeability and effective porosity of porous media. Phys. Rev. E **56**, 3319 (1997)
20. A. Koponen, M. Kataja, J. Timonen, Tortuous flow in porous media. Phys. Rev. E **54**, 406 (1996)
21. M.A. Knackstedt, X. Zhang, Direct evaluation of length scales and structural parameters associated with flow in porous media. Phys. Rev. E **50**, 2134 (1994)
22. M. Matyka, A. Khalili, Z. Koza, Tortuosity-porosity relation in porous media flow. Phys. Rev. E **78**, 026306 (2008)
23. Z. Koza, M. Matyka, A. Khalili, Finite-size anisotropy in statistically uniform porous media. Phys. Rev. E **79**, 066306 (2009)
24. J. Bear, *Dynamics of Fluid in Porous Media* (New York, 1972)
25. R. Nemati, J. Rahbar Shahrouzi, R. Alizadeh, A stochastic approach for predicting tortuosity in porous media via pore network modeling. Comput. Geotech. **120**, 103406 (2020)
26. A. Ghassemi, A. Pak, Pore scale study of permeability and tortuosity for flow through particulate media using lattice Boltzmann method. Int. J. Numer. Anal. Methods Geomech. **35**, 886 (2011)
27. D. Froning, J. Yu, G. Gaiselmann, U. Reimer, I. Manke, V. Schmidt, W. Lehnert, Impact of compression on gas transport in non-woven gas diffusion layers of high temperature polymer electrolyte fuel cells. J. Power Sources **318**, 26 (2016)
28. J. Sarkar, S. Bhattacharyya, Application of graphene and graphene-based materials in clean energy-related devices Minghui. Arch. Thermodyn. **33**, 23 (2012)
29. H. Saomoto, J. Katagiri, Direct comparison of hydraulic tortuosity and electric tortuosity based on finite element analysis. Theor. Appl. Mech. Lett. **5**, 177 (2015)
30. H. Saomoto, J. Katagiri, Particle shape effects on hydraulic and electric tortuosities: a novel empirical tortuosity model based on van Genuchten-type function. Transp. Porous Media **107**, 781 (2015)
31. B. Sheikh, A. Pak, Numerical investigation of the effects of porosity and tortuosity on soil permeability using coupled three-dimensional discrete-element method and lattice Boltzmann method. Phys. Rev. E Stat. Nonlin. Soft Matter Phys. **91**, 1 (2015)
32. Y. Jin, J.B. Dong, X. Li, Y. Wu, Kinematical measurement of hydraulic tortuosity of fluid flow in porous media. Int. J. Mod. Phys. C **26** (2015)
33. H. Juliust, M.N. Amien, G.T. Pantouw, F.D. Eljabbar Latief, Complexity analysis of interconnected pore using hydraulic tortuosity. IOP Conf. Ser. Earth Environ. Sci. **311**, 012034 (2019)
34. A. Nabovati, A.C.M. Sousa, Fluid flow simulation in random porous media at pore level using the lattice Boltzmann method. Sci. Technol. **2**, 226 (2007)
35. E.E. Petersen, Diffusion in a pore of varying cross section. AIChE J. **4**, 343 (1958)
36. J. Van Brakel, P.M. Heertjes, Analysis of diffusion in macroporous media in terms of a porosity, a tortuosity and a constrictivity factor. Int. J. Heat Mass Transf. 1093 (1974)
37. L. Holzer et al., Fundamental relationships between 3D pore topology, electrolyte conduction and flow properties: towards knowledge-based design of ceramic diaphragms for sensor applications. Mater. Des. **99**, 314 (2016)
38. L. Holzer, D. Wiedenmann, B. Münch, L. Keller, M. Prestat, P. Gasser, I. Robertson, B. Grobéty, The influence of constrictivity on the effective transport properties of porous layers in electrolysis and fuel cells. J. Mater. Sci. **48**, 2934 (2013)

39. C.F. Berg, Permeability description by characteristic length, tortuosity, constriction and porosity. Transp. Porous Media **103**, 381 (2014)
40. C.F. Berg, Re-examining Archie's law: conductance description by tortuosity and constriction. Phys. Rev. E **86**, 046314 (2012)
41. O. Stenzel, O. Pecho, L. Holzer, M. Neumann, V. Schmidt, Predicting effective conductivities based on geometric microstructure characteristics. AIChE J. **62**, 1834 (2016)
42. D. Wiedenmann et al., Three-dimensional pore structure and ion conductivity of porous ceramic diaphragms. AIChE J. **59**, 1446 (2013)
43. C.F. Berg, W.D. Kennedy, D.C. Herrick, Conductivity in partially saturated porous media described by porosity, electrolyte saturation and saturation-dependent tortuosity and constriction factor. Geophys. Prospect. **70**, 400 (2022)
44. M. Neumann, C. Hirsch, J. Staněk, V. Beneš, V. Schmidt, Estimation of geodesic tortuosity and constrictivity in stationary random closed sets. Scand. J. Stat. **46**, 848 (2019)
45. B. Münch, L. Holzer, Contradicting geometrical concepts in pore size analysis attained with electron microscopy and Mercury intrusion. J. Am. Ceram. Soc. **91**, 4059 (2008)
46. G. Matheron, *Random Sets and Integral Geometry* (J. Wiley & Sons, New York, 1975)
47. G.E. Archie, The electrical resistivity log as an aid in determining some reservoir characteristics. Trans. AIME **146** (1942)
48. M.R.J. Wyllie, W.D. Rose, Some theoretical considerations related to the quantitative evaluation of the physical characteristics of reservoir rock from electrical log data. J. Petrol. Technol. **2**, 105 (1950)
49. A. Wiegmann, A. Zemitis, EJ-HEAT: a fast explicit jump harmonic averaging solver for the effective heat conductivity of composite materials (2006)
50. S.J. Cooper, A. Bertei, P.R. Shearing, J.A. Kilner, N.P. Brandon, TauFactor: an open-source application for calculating tortuosity factors from tomographic data. SoftwareX **5**, 203 (2016)
51. J.E. Owen, The resistivity of a fluid-filled porous body. J. Petrol. Technol. **4**, 169 (1952)
52. F.A.L. Dullien, *Porous Media: Fluid Transport and Pore Structure* (Academic Press Ltd., London, 2012)
53. J. He, Y. Ju, L. Lammers, K. Kulasinski, L. Zheng, Tortuosity of kerogen pore structure to gas diffusion at molecular- and nano-scales: a molecular dynamics simulation. Chem. Eng. Sci. **215**, 115460 (2020)
54. S.G. Jennings, The mean free path in air. J. Aerosol. Sci. **19**, 159 (1988)
55. W.G. Pollard, R.D. Present, On gaseous self-diffusion in long capillary tubes. Phys. Rev. **73**, 762 (1948)
56. J.W. Veldsink, R.M.J. van Damme, G.F. Versteeg, W.P.M. van Swaaij, The use of the dusty-gas model for the description of mass transport with chemical reaction in porous media. Chem. Eng. J. Biochem. Eng. J. **57**, 115 (1995)
57. C.N. Satterfield, Th.K. Sherwood, *The Role of Diffusion in Catalysis* (Addison-Wesley Pub. Co., 1963)
58. F. Elwinger, P. Pourmand, I. Furó, Diffusive transport in pores. Tortuosity and molecular interaction with the pore wall. J. Phys. Chem. C **121**, 13757 (2017)
59. J.A. Currie, Gaseous diffusion in porous media. Part 2.—dry granular materials. Br. J. Appl. Phys. **11**, 318 (1960)
60. A.A. Garrouch, L. Ali, F. Qasem, Using diffusion and electrical measurements to assess tortuosity of porous media. Ind. Eng. Chem. Res. **40**, 4363 (2001)
61. L.J. Klinkenberg, Analogy between diffusion and electrical conductivity in porous rocks. GSA Bull. **62**, 559 (1951)
62. P. Wong, Conductivity, permeability and electrokinetics, in *Methods in the Physics of Porous Media*. ed. by P. Wong (Academic Press Ltd., London, 1999), pp.115–159
63. H. Iwai et al., Quantification of SOFC anode microstructure based on dual beam FIB-SEM technique. J. Power Sources **195**, 955 (2010)
64. F. Tariq, V. Yufit, M. Kishimoto, P.R. Shearing, S. Menkin, D. Golodnitsky, J. Gelb, E. Peled, N.P. Brandon, Three-dimensional high resolution X-ray imaging and quantification of lithium ion battery mesocarbon microbead anodes. J. Power Sources **248**, 1014 (2014)

65. B. Tjaden, D.J.L. Brett, P.R. Shearing, Tortuosity in electrochemical devices: a review of calculation approaches. Int. Mater. Rev. **63**, 47 (2018)
66. J. Fu, H.R. Thomas, C. Li, Tortuosity of porous media: image analysis and physical simulation. Earth Sci. Rev. 1 (2020)
67. T.G. Tranter, M.D.R. Kok, M. Lam, J.T. Gostick, Pytrax: a simple and efficient random walk implementation for calculating the directional tortuosity of images. SoftwareX **10**, 100277 (2019)
68. J. Latt et al., Palabos: parallel lattice Boltzmann solver. Comput. Math. Appl. **81**, 334 (2021)
69. H. Babovsky, On Knudsen flows within thin tubes. J. Stat. Phys. **44**, 865 (1986)
70. J.M. Zalc, S.C. Reyes, E. Iglesia, The effects of diffusion mechanism and void structure on transport rates and tortuosity factors in complex porous structures. Chem. Eng. Sci. **59**, 2947 (2004)
71. J.C. Ferguson, A. Borner, F. Panerai, S. Close, N.N. Mansour, Continuum to rarefied diffusive tortuosity factors in porous media from X-ray microtomography. Comput. Mater Sci. **203**, 111030 (2022)
72. X. Gao, J.C. Diniz da Costa, S.K. Bhatia, Understanding the diffusional tortuosity of porous materials: an effective medium theory perspective. Chem. Eng. Sci. **110**, 55 (2014)
73. L. Holzer, B. Iwanschitz, Th. Hocker, L. Keller, O. Pecho, G. Sartoris, Ph. Gasser, B. Muench, Redox cycling of Ni–YSZ anodes for solid oxide fuel cells: influence of tortuosity, constriction and percolation factors on the effective transport properties. J. Power Sources **242**, 179 (2013)
74. P. Soille, *Morphological Image Analysis* (Springer, Berlin Heidelberg, 2004)
75. D. Jeulin, L. Vincent, G. Serpe, Propagation algorithms on graphs for physical applications. J. Vis. Commun. Image Represent. **3**, 161 (1992)
76. L.M. Keller, L. Holzer, R. Wepf, P. Gasser, 3D geometry and topology of pore pathways in Opalinus clay: implications for mass transport. Appl. Clay Sci. **52**, 85 (2011)
77. L.M. Keller, L. Holzer, R. Wepf, P. Gasser, B. Münch, P. Marschall, On the application of focused ion beam nanotomography in characterizing the 3D pore space geometry of Opalinus clay. Phys. Chem. Earth **36** (2011)
78. W.B. Lindquist, S.-M. Lee, D.A. Coker, K.W. Jones, P. Spanne, Medial axis analysis of void structure in three-dimensional tomographic images of porous media. J. Geophys. Res. Solid Earth **101**, 8297 (1996)
79. W. Dijkstra, A note on two problems in connection with graphs. Numer. Math. (Heidelb) **1**, 269 (1959)
80. M. Sato, I. Bitter, M.A. Bender, A.E. Kaufman, M. Nakajiama, *TEASAR: tree-structure extraction algorithm for accurate and robust skeletons*, in Eighth Pacific Conference on Computer Graphics and Applications (PG''00) (2000), p. 281.
81. S.J. Cooper et al., Image based modelling of microstructural heterogeneity in LiFePO$_4$ electrodes for Li-ion batteries. J. Power Sources **247**, 1033 (2014)
82. G. Gaiselmann, M. Neumann, V. Schmidt, O. Pecho, T. Hocker, L. Holzer, Quantitative relationships between microstructure and effective transport properties based on virtual materials testing. AIChE J. **60**, 1983 (2014)
83. R.I. Al-Raoush, I.T. Madhoun, TORT3D: a MATLAB code to compute geometric tortuosity from 3D images of unconsolidated porous media. Powder Technol. **320**, 99 (2017)
84. R. Thiedmann, C. Hartnig, I. Manke, V. Schmidt, W. Lehnert, Local structural characteristics of pore space in GDLs of PEM fuel cells based on geometric 3D graphs. J. Electrochem. Soc. **156**, B1339 (2009)
85. D. Jungnickel, *Graphs, Network and Algorithm* (Springer, Berlin, 1999)
86. W. Sobieski, The use of path tracking method for determining the tortuosity field in a porous bed. Granul. Matter **18**, 1 (2016)
87. W. Sobieski, M. Matyka, J. Gołembiewski, S. Lipiński, The path tracking method as an alternative for tortuosity determination in granular beds. Granul. Matter **20**, (2018)
88. W. Sobieski, Numerical investigations of tortuosity in randomly generated pore structures. Math. Comput. Simul. **166**, 1 (2019)

89. C. Lantuejoul, S. Beucher, On the use of the geodesic metric in image analysis. J. Microsc. **121**, 39 (1981)
90. C.J. Gommes, A.-J. Bons, S. Blacher, J.H. Dunsmuir, A.H. Tsou, Practical methods for measuring the tortuosity of porous materials from binary or gray-tone tomographic reconstructions. AIChE J. **55**, 2000 (2009)
91. E. Machado Charry, M. Neumann, J. Lahti, R. Schennach, V. Schmidt, K. Zojer, Pore space extraction and characterization of sack paper using μ-CT. J. Microsc **272**, 35 (2018)
92. K. Thulasiraman, M.N.S. Swamy, *Graphs, Theory and Algorithms* (John Wiley & Sons, New York, 1992)
93. S.N. Chiu, D. Stoyan, W. Kendall, J. Mecke, *Stochastic Geometry and Its Applications*, 3rd edn. (John Wiley & Sons, Chichester, UK, 2013)
94. M. Neumann, O. Stenzel, F. Willot, L. Holzer, V. Schmidt, Quantifying the influence of microstructure on effective conductivity and permeability: virtual materials testing. Int. J. Solids Struct. **184**, 211 (2020)
95. J. Vicente, F. Topin, J.V. Daurelle, Open celled material structural properties measurement: from morphology to transport properties. Mater. Trans. **47**, 2195 (2006)
96. J.A. Sethian, A fast marching level set method for monotonically advancing fronts. Proc. Natl. Acad. Sci. U.S.A. **93**, 1591 (1996)
97. J.A. Sethian, Fast marching methods. SIAM Rev. **41**, 199 (1999)
98. J.A. Sethian, Evolution, implementation, and application of level set and fast marching methods for advancing fronts. J. Comput. Phys. **169**, 503 (2001)
99. J.A. Sethian, A.M. Popovici, 3-D imaging using higher order fast marching traveltimes. Geophysics **64**, 516 (2002)
100. P.S. Jørgensen, S.L. Ebbehøj, A. Hauch, Triple phase boundary specific pathway analysis for quantitative characterization of solid oxide cell electrode microstructure. J. Power Sources **279**, 686 (2015)
101. P.S. Jørgensen, K.V. Hansen, R. Larsen, J.R. Bowen, Geometrical characterization of interconnected phase networks in three dimensions. J. Microsc **244**, 45 (2011)
102. O.O. Taiwo, D.P. Finegan, D.S. Eastwood, J.L. Fife, L.D. Brown, J.A. Darr, P.D. Lee, D.J.L. Brett, P.R. Shearing, Comparison of three-dimensional analysis and stereological techniques for quantifying Lithium-ion battery electrode microstructures. J. Microsc. **263**, 280 (2016)
103. T. Hamann, L. Zhang, Y. Gong, G. Godbey, J. Gritton, D. McOwen, G. Hitz, E. Wachsman, The effects of constriction factor and geometric tortuosity on Li-ion transport in porous solid-state Li-ion electrolytes. Adv. Funct. Mater. **30**, (2020)
104. M. Neumann, E. Machado Charry, K. Zojer, V. Schmidt, On variability and interdependence of local porosity and local tortuosity in porous materials: a case study for sack paper. Methodol. Comput. Appl. Probab. **23**, 613 (2021)
105. Y.C.K. Chen-Wiegart, R. Demike, C. Erdonmez, K. Thornton, S.A. Barnett, J. Wang, Tortuosity characterization of 3D microstructure at nano-scale for energy storage and conversion materials. J. Power Sources **249**, 349 (2014)
106. A. Çeçen, E.A. Wargo, A.C. Hanna, D.M. Turner, S.R. Kalidindi, E.C. Kumbur, 3-D Microstructure analysis of fuel cell materials: spatial distributions of tortuosity, void size and diffusivity. J. Electrochem. Soc. **159**, B299 (2012)
107. P.R. Shearing, L.E. Howard, P.S. Jørgensen, N.P. Brandon, S.J. Harris, Characterization of the 3-dimensional microstructure of a graphite negative electrode from a Li-ion battery. Electrochem. Commun. **12**, 374 (2010)
108. F.L.E. Usseglio-Viretta et al., Resolving the discrepancy in tortuosity factor estimation for Li-ion battery electrodes through micro-macro modeling and experiment. J. Electrochem. Soc. **165**, A3403 (2018)

Chapter 3
Tortuosity-Porosity Relationships: Review of Empirical Data from Literature

Abstract It is generally assumed that transport resistance in porous media, which can also be expressed as tortuosity, correlates somehow with the pore volume fraction. Hence, mathematical expressions such as the Bruggeman relation (i.e., $\tau^2 = \varepsilon^{-1/2}$) are often used to describe tortuosity (τ)—porosity (ε) relationships in porous materials. In this chapter, the validity of such mathematical expressions is critically evaluated based on empirical data from literature. More than 2200 datapoints (i.e., $\tau - \varepsilon$ couples) are collected from 69 studies on porous media transport. When the empirical data is analysed separately for different material types (e.g., for battery electrodes, SOFC electrodes, sandstones, packed spheres etc.), the resulting τ versus ε—plots do not show clear trend lines, that could be expressed with a mathematical expression. Instead, the datapoints for different materials show strongly scattered distributions in rather ill-defined 'characteristic' fields. Overall, those characteristic fields are strongly overlapping, which means that the $\tau - \varepsilon$ characteristics of different materials cannot be separated clearly. When the empirical data is analysed for different tortuosity types, a much more consistent pattern becomes apparent. Hence, the observed $\tau - \varepsilon$ pattern indicates that the measured tortuosity values strongly depend on the involved type of tortuosity. A relative order of measured tortuosity values then becomes apparent. For example, the values observed for direct geometric and mixed tortuosities are concentrated in a relatively narrow band close to the Bruggeman trend line, with values that are typically < 2. In contrast, indirect tortuosities show higher values, and they scatter over a much larger range. Based on the analysis of empirical data, a detailed pattern with a very consistent relative order among the different tortuosity types can be established. The main conclusion from this chapter is thus that the tortuosity value that is measured for a specific material, is much more dependent on the type of tortuosity than it is dependent on the material and its microstructure. The empirical data also illustrates that tortuosity is not strictly bound to porosity. As the pore volume decreases, the more scattering of tortuosity values can be observed. Consequently, any mathematical expression that aims to

Supplementary Information The online version contains supplementary material available at https://doi.org/10.1007/978-3-031-30477-4_3.

© The Author(s) 2023

L. Holzer et al., *Tortuosity and Microstructure Effects in Porous Media*,
Springer Series in Materials Science 333,
https://doi.org/10.1007/978-3-031-30477-4_3

provide a generalized description of $\tau - \varepsilon$ relationships in porous media must be questioned. A short section is thus provided with a discussion of the limitations of such mathematical expressions for $\tau - \varepsilon$ relationships. This discussion also includes a description of the rare and special cases, for which the use of such mathematical expressions can be justified.

3.1 Introduction

The review of concepts and theories (see Chap. 2) reveals a multitude of different approaches, how tortuosity can be defined and measured nowadays. This diversity raises the question, whether the different tortuosity types can be used interchangeably, which is, however, only the case when they reveal identical or very similar results when applied to the same sample and microstructure. If it turns out, that different tortuosity types reveal different results, then the question arises, whether these differences are systematic, predictable, and understandable. A deeper understanding of inherent differences will improve our interpretation of measured tortuosity data and sharpen the scientific discussion of the topic.

To address these issues, we compile and analyse empirical data from the literature in this chapter. Table 3.1 represents a list of data sets with tortuosity-porosity ($\tau - \varepsilon$) couples collected in almost 70 different studies. Each line in this table represents a set of data points for a specific class of porous material, which is characterized with a specific tortuosity type. More than 2200 data points were collected semi-automatically from literature (using the webplotdigitizer [70]), in order to investigate and identify specific $\tau-\varepsilon$ patterns.

$\tau-\varepsilon$ relationships are collected for the following 7 important *classes of porous materials*:

(a) Solid oxide fuel cell (SOFC) electrodes and sintered ceramics,
(b) Gas diffusion layers (GDL) of polymer electrolyte membrane fuel cells (PEM-FC) and other fibrous materials (e.g., paper, particle filters),
(c) Battery electrodes, mostly from Li-ion batteries (LIB),
(d) Geological materials (sandstones, clays, soils),
(e) 2D-models of granular materials (packed circles, ellipses, squares, or rectangles)
(f) 3D-models of granular materials (packed spheres and ellipsoids, mono-sized and poly-dispersed, and experimental model materials (e.g., glass beads)
(g) 3D-models of networked pore structures (from stochastic simulation) and foams.

In Sect. 3.2 we analyse this data to see, whether different materials classes reveal specific $\tau-\varepsilon$ patterns. In our comparison of empirical data, we also consider the different tortuosity types as a criterion for discrimination. In Sect. 3.3, the data from the seven material classes is thus also used to investigate the impact of different tortuosity types on the observed $\tau-\varepsilon$ patterns. Thereby, we use the classification scheme from Chap. 2 with the main categories: (I) direct geometric τ, (II) mixed τ and (III) indirect physics-based tortuosities (see also Fig. 2.8). In Sect. 3.4, we consider some specific examples with datasets, where different tortuosity types are applied to the very same microstructure. From these investigations, a consistent pattern can be deduced, which shows how various tortuosity types differ from each other (Sect. 3.5). Finally, in Sect. 3.6, the mathematical descriptions of τ–ε relationships in literature (e.g., Bruggeman, Archie, Maxwell etc.) are reviewed in the light of the findings in previous sections.

3.2 Empirical Data for Different Materials and Microstructure Types

The empirical data in tables 3.1 covers a large variety of materials with a wide range of porosities and with different microstructure characteristics. The data includes, for example, simple structures consisting of mono-sized spheres, but also more complex structures of packed fibers, foams, and composite materials (e.g., SOFC electrodes).

The empirical tortuosity-porosity (τ–ε) data is plotted in Fig. 3.1a–g for each specific material type separately. Overall, these plots show a relatively large scatter, so that no specific τ–ε relationship that could be described with a simple trend line and/or a mathematical expression (such as $\tau = \varepsilon^x$) can be attributed to the specific material types. This finding indicates that even within a single material type the structural variety is large. However, when taking a closer look at Fig. 3.1a–g, for each material type a characteristic field (i.e., a dense cloud of data points) can be observed in the τ–ε plots.

Figure 3.1h illustrates that the characteristic fields of the different material types tend to have a strong overlap. Hence, it seems that τ–ε relationships are not suitable to capture the strong differences between the involved types of materials and structures. Nevertheless, for simple microstructures, such as packed spheres and ellipsoids, the scatter of characteristic fields is relatively small (see Fig. 3.2 e, f). For these rather simple microstructures, the corresponding data points are in a narrow band close to and often slightly above the Bruggeman trend line (i.e., $T = \tau^2 = 1/\varepsilon^{0.5}$). For all other types of materials and microstructures, the tortuosity-porosity (τ–ε) data shows as strong scattering.

It must be emphasized that the plots in Fig. 3.1 do not distinguish between different types of tortuosities, which may be one reason for the observed scatter of τ–ε datapoints.

The underlying source file for Fig. 3.1, with detailed information to 69 references, can be downloaded from the electronic appendix (Supplementary File 3.1).

Table 3.1 a–g contains list with empirical tortuosity-porosity (τ–ε)-data from literature 2204 tortuosity-porosity (τ–ε) data points are collected from 69 references in literature. The corresponding source file for Table 3.1, which contains additional detailed information, can be downloaded from the electronic appendix (Supplementary File 3.1). It must be emphasized that 'porosity' in this table always means the effective porosity, where this information is available. The effective porosity represents the fraction of pore space, which forms a contiguous network, and which excludes trapped pores. In subsequent figures derived from this table the effective porosity is plotted.

Table 3.1a SOFC electrodes and porous ceramics (e.g., separation membrane for SOEC, ceramic catalyst support, or sintered alumina) *Legend:* (Nr of dataset) Referen*ce [Ref Nr],* τ *type,* (τ methodical details) investigated material

(1)	Grew et al. [1], τ_{indir_diff}, (from sim with Laplace solver, Fick's diff) Ni-YSZ anode
(2)	Grew et al. [2], τ_{indir_diff}, (from sim with Laplace solver, Fick's diff) Ni-YSZ anode
(3)	Iwai et al. [3], τ_{indir_diff}, (from sim with random walk, bulk diff) Ni-YSZ anode
(4)	Kishimoto et al. [4], τ_{indir_diff}, (from sim with LBM, bulk diff) Ni-YSZ anode
(5)	Cooper et al. [5], τ_{indir_diff}, (CFD-FVM, Avizo, random walk, 'StarCCM+') LSCF
(6)	Cooper et al. [6], comparison of various τ-types specified below, LSCF cathode
(6a)	$\tau_{dir\_pore\_centroid}$, (image analysis/Avizo Fire, voxel based)
(6b)	$\tau_{indir\_diff\_sim}$. (sim conductivity, 'StarCCM+'/Laplace s., Fourier's law, mesh based)
(6c)	$\tau_{indir\_diff\_sim}$. (sim bulk diffusion, 'AvizoXlab'/Laplace solver, Fick's law, voxel based)
(6d)	$\tau_{indir\_diff\_sim}$, (simbulk diffusion, 'TauFactor'/Laplace solver, Fick's law, voxel based)
(6e)	$\tau_{indir\_diff\_sim}$. (simulation of bulk diffusion, (in-house)/random walk, voxel based)
(7)	Wilson et al. [7], τ_{indir_diff}, (sim with Laplace solver, Fick's diff) Ni-YSZ anode
(8)	Tjaden et al. [8], comparison of tortuosity types, porous YSZ support layer:
(8a)	τ_{dir_FMM}, (image analysis/in-house (Matlab)/FIB-SEM)
(8b)	τ_{dir_FMM}, (image analysis/in-house (Matlab)/X-ray nano CT)
(8c)	τ_{indir_therm}, (sim. of thermal cond., 'StarCCM+'/Laplace, Fourier, mesh based/ FIB-SEM)
(8d)	τ_{indir_therm}, (sim. of thermal cond., 'StarCCM+'/Laplace, Fourier, mesh based/XnCT)
(8e)	τ_{indir_diff}, (diffusion cell experiment/Gas diffusion at 30 and 100 °C/Fick's law)
(9)	Joos et al. [9], τ_{indir_diff}, (from simulation with Laplace solver) LSCF cathode
(10)	Laurencin et al. [10], τ_{indir_diff}, (from sim with Laplace solver) Ni-YSZ anode
(11)	Holzer et al. [11], $\tau_{dir\_medial\_axis}$, (Avizo skeletonization/Matlab) Ni-YSZ anode
(12a)	Pecho et al. [12], $\tau_{dir\_geodesic}$, (Image analysis, in-house) Ni-YSZ anode
(12b)	Pecho et al. [12], supplementary material
(13)	Holzer et al. [13], $\tau_{dir\_geodesic}$, (Image analysis, in-house) Ni-YSZ anode
(14a)	Wiedenmann et al. [14], $\tau_{dir\_medial\_axis}$, (Avizo/Matlab) sintered Olivine/Wollastonite
(14b)	Wiedenmann et al. [14], $\tau_{indir\_ele\_exp}$, (from EIS experiment) Olivine/Wollastonite
(15)	Zheng et al. [15], τ_{indir_diff}, (from sim with Laplace solver) stochastic model for anode

(continued)

Table 3.1   (continued)

(16)	Lichtner et al. [16], τ_{indir_diff}, (from sim with Laplace solver, GeoDict) LSM-YSZ cathode
(17)	Almar et al. [17], τ_{indir_diff}, (from sim with Laplace solver) LSCF and BSCF cathodes
(18)	Endler et al. [18], τ_{indir_diff}, (from sim with Laplace solver) LSCF cathode
(19)	Holzer et al. [19], 2 tortuosity types, porous Zr-oxide used as Diaphragm in pH sensor:
(19a)	$\tau_{dir\_geodesic}$, (image analysis/in-house)
(19b)	τ_{indir_ele}, (simulation of electrical conduction, GeoDict/Laplace solver/Ohm's law)
(20)	Haj et al. [20], $\tau_{dir\_geodesic}$, (Image analysis, in-house) sintered Ni
(21a)	Kishimoto et al. [21], τ_{indir_diff}, (Laplace solver, Fick's diff) pores in CGO anode
(21b)	Kishimoto et al. [21], τ_{indir_ele}, (Laplace solver, Ohm's law) solid phase in CGO anode
(22a)	Shanti et al. [22], τ_{indir_diff}, (from sim with Laplace solver, gas diff), sintered alumina
(22b)	Shanti et al. [22], $\tau_{dir\_skeleton}$, (Amira skeletonization/shortest path analysis), alumina

Table 3.1b   PEM FC (GDL) and fibrous materials. *Legend:* (Nr of dataset) Reference *[Ref Nr]*, τ *type*, (τ methodical details) investigated material

(23)	Yu et al. [23], $\tau_{indir\_diff\_exp}$, (from diffusion experiment) Pt electrode
(24)	Sarkar and Bhattacharyya [24], $\tau_{mixed\_hydr\_Vav}$, (from LBM-sim, Navier–Stokes) GDL through-plane dir
(25a)	Garcia-Salaberry et al. [25], τ_{indir_diff}, (from LBM-sim) GDL through-plane, var. thicknesses
(25b)	Garcia-Salaberry et al. [25], τ_{indir_diff}, (from LBM-sim) GDL in-plane, various thicknesses
(26)	Froning et al. [26], $\tau_{mixed\_hydr\_Vav}$, (from LBM-sim, NS gas flow) GDL in-plane direction
(27a)	Flückiger et al. [27], $\tau_{indir\_diff\_exp}$, (from experiment) GDL dry, no Filler
(27b)	Flückiger et al. [27], $\tau_{indir\_diff\_exp}$, (from experiment) GDL dry, with Filler
(28)	Froning, et al. [28], $\tau_{mixed\_hydr\_Vav}$, (from LBM, NS) PEM GDL real/virtual, w/o Filler
(29)	Holzer et al. [29], 2 tortuosity types, PEM GDL dry IP and TP, compression series:
(29a)	$\tau_{dir\_geodesic}$, (image analysis, in-house) var. thickness from in-situ μ-CT compression experiment
(29b)	τ_{indir_ele} (simulation of electrical conduction, GeoDict/Laplace solver/Ohm's law)
(30)	Holzer et al. [30], 2 tortuosity types, PEM GDL wet IP and TP, μ-CT imbibition experiment:
(30a)	$\tau_{dir\_geodesic}$, (image analysis/in-house, from dynamic XCT-imbibition experiment)
(30b)	τ_{indir_ele}, (simulation of electrical conduction, GeoDict/Laplace solver/Ohm's law)
(31)	Huang et al. [31], $\tau_{dir\,FMM}$, (from image analysis) fibrous cloths

Table 3.1c Battery electrodes. *Legend:* (Nr of dataset) Reference *[Ref Nr]*, τ *type,* (τ methodical details) investigated material

(32)	Cooper et al. [32], LI-Battery (LiFePO$_4$):
(32a)	$\tau_{dir\_pore\_centroid}$, (image analysis/Avizo Fire, voxel based)
(32b)	τ_{indir_therm}, (sim. thermal cond., 'StarCCM+'/Laplace solver Fourier's law, mesh based)
(33)	Tariq et al. [33], τ_{indir_diff}, (Star-CCM+, CD adapco, FVM) LIB (graphite)
(34)	Shearing et al. [34], τ_{indir_diff}, (Star-CCM + , CD adapco, FVM) LIB (LMO)
(35)	Taiwo et al. [35], τ_{dir_FMM}, (image analysis, in-house SW) LIB (LCO, LMO, graphite)
(36)	Almar et al. [36], τ_{indir_diff}, (GeoDict, Laplace solver) LIB
(37)	Shearing et al. [37], τ_{indir_diff}, (TauFactor, Laplace solver) LIB (graphite)
(38)	Hutzenlaub et al. [38], τ_{indir_diff}, (GeoDict, Laplace solver) LIB
(39)	Ebner et al. [39], τ_{indir_diff}, (Bruggeman Est., Laplace) LIB (LCO, LMO, graphite)
(40)	Hamann et al. [40], τ_{dir_FMM}, (image analysis, Matlab, Fiji) LLCZN garnet electrolyte
(41)	Landesfeind et al. [41], $\tau_{indir\_ele\_exp}$, (EIS experiment) LIB (3–6% binder)
(42a)	Landesfeind et al. [42], $\tau_{indir\_ele\_exp}$, (EIS experiment) LIB (NMC, NCA, graphite)
(42b)	Landesfeind et al. [42], τ_{indir_ele}, (Laplace solver) LIB (NMC, NCA, graphite)
(43)	Various authors, cited in Landesfeind et al. [43], τ_{indir_ele}, LIB

Table 3.1d Earth Science and Geo-engineering (sandstones, clays/soils/mortar). *Legend:* (Nr of dataset) Referenc*e [Ref Nr]*, τ *type,* (τ methodical details) investigated material

(44)	Berg [44], $\tau_{dir\_skeleton}$ (flux-weighted, e-Core network model) Fontainebleau sandstone
(45)	Provis et al. [45], τ_{indir_diff}, (Random walk sim., in-house) Alkali activated fly ash mortar
(46a)	Klinkenberg [46], cited in Ziarani [47] τ_{indir_ele}, (sim. Laplace solver) soil
(46b)	Klinkenberg [46, 47] τ_{indir_ele}, (Laplace) unconsolidated sand and glass beads
(46c)	Klinkenberg [46, 47] τ_{indir_ele}, (Laplace) sandstones and limestones
(48)	Keller et al. [48], $\tau_{dir\_medial\_axis}$, (Avizo skeletonization/Matlab) Opalinus Clay
(49)	Berg [49], $\tau_{dir\_skeleton}$ (flux-weighted, e-Core network model) Bentheimer sandstone
(50)	Keller et al. [50]$\tau_{dir\_medial\_axis}$, (Avizo skeletonization/Matlab) Opalinus Clay
(51)	Various authors cited in Kristensen et al. [51], $\tau_{indir\_diff\_exp}$, (experiments) soils
(52)	Boving et al. [52] (cited in [51]), $\tau_{indir\_diff\_exp}$, (tracer diffusion experiments) soils
(53)	Nemati [53], $\tau_{dir\_skeleton}$ (flux-weighted, Pore network model) Berea sandstone

Table 3.1e Granular porous media I: 2D models (circles, ellipsoids, rectangles). *Legend:* (Nr of dataset) Referenc*e [Ref Nr]*, τ *type,* (τ methodical details) investigated material

(54a)	Saomoto and Katagiri [54], $\tau_{mixed\_hydr\_Streamline}$, (Comsol + image analysis) 2D monosized
(54b)	Saomoto and Katagiri [54], $\tau_{mixed\_ele\_Streamline}$, (Comsol + image analysis) 2D monosized

(continued)

Table 3.1e (continued)

(55)	Saomoto et al. [55], mixed tortuosity types, monosized 2D ellipsoids, aspect ratios 1- 5
(55a)	$\tau_{mixed\_hydr\_Streamline}$, (sim. of flow with Comsol/Image Analysis (IA) of 2D vel. field)
(55b)	$\tau_{mixed\_ele\_Streamline}$, (simulation of el. conduction with Comsol/IA of 2D velocity field)
(55c)	$\tau_{mixed\_hydr\_Vav}$, (simulation of flow with Comsol/IA of 2D velocity field)
(55d)	$\tau_{mixed\_ele\_Vav}$, (simulation of el. conduction with Comsol/IA of 2D velocity field)
(56)	Nabovati et al. [56], $\tau_{mixed\_hydr\_Vav}$, (LBM, Navier Stokes) 2D squares and rectangles
(57)	Ghassemi and Pak [57], $\tau_{mixed\_diff\_Vav}$, (LBM) 2D spheres polydispersed

Table 3.1f Granular porous media II: 3D models (packed spheres, ellipsoids, sands) and model materials (glass beads, sand). *Legend:* (Nr of dataset) Reference *[Ref Nr]*, τ *type*, (τ methodical details) investigated material

(39)	Ebner et al. [39], τ_{indir_diff}, (Bruggeman Estimator/Laplace) sphere packing model
(58)	Gommes et al. [58], $\tau_{dir\_geodesic}$, (Matlab, in-house) Battery Poisson sphere model
(59)	Chung et al. [59], τ_{indir_diff}, (Batt3D/Laplace) Battery sphere model with ESyS
(60)	Sobieski [60], τ_{dir_PTM} (in-house, virtual 3D with DEM) polydispersed spheres
(61a)	Sheikh and Pak [61], $\tau_{mixed\_diff\_Vav}$ (LBM C++, IA) 3D polydispersed spheres
(61b)	Sheikh and Pak [61], τ_{indir_diff} (LBM C++) polydispersed spheres with DEM
(62)	Pawlowski et al. [62], $\tau_{mixed\_hydr\_SL}$ (Open Foam/VGSMax/ImageVis3D) chromatogr.
(63)	Al-Roush and Madhoun [63], $\tau_{dir\_medial\_axis}$ (Tort3D/Matlab) Silica Sand polydisp.
(64a)	Hormann et al. [64], $\tau_{dir\_geodesic}$ (Fiji, in-house), Silica monoliths/chromatogr.
(64b)	Hormann et al. [64], $\tau_{dir\_medial\_axis}$ (Fiji, skeletonize3D/analyzeSkeleton) Silica
(46)	Klinkenberg [46], in Ziarani [47, 61], $\tau_{indir\_ele\_exp}$, non-consolidated sand
(65)	Johnson et al. [65], τ_{indir_diff} (TauFactor) packed agar/cellulose for chromatogr.

Table 3.1g Foams and stochastic models of networked structures with continuous phases. *Legend:* (Nr of dataset) Reference *[Ref Nr]*, τ *type*, (τ methodical details) investigated material

(66a)	Stenzel et al. [66], $\tau_{dir\_geodesic}$ (in-house) SSGM stochastic spatial graph model
(66b)	Stenzel et al. [66], τ_{indir_ele} (GeoDict, Laplace solver) SSGM graph model, contin. phase
(67)	Gaiselmann et al. [67], $\tau_{dir\_medial\_axis}$, (Avizo skeleton/Matlab) SSGM graph model
(68)	Knackstedt et al. [68], τ_{indir_diff} (Laplace, bulk diffusion) PU open cell foams
(69)	Vicente et al. [69], τ_{dir_FMM} (in-house image analysis) Metal open cell foams
(22a)	Shanti et al. [22], $\tau_{dir\_medial\_axis}$ (Image analysis) Alumina packed spheres polydispersed
(22b)	Shanti et al. [22], τ_{indir_diff} (Laplace, bulk diffusion) Alumina packed spheres polydisp.

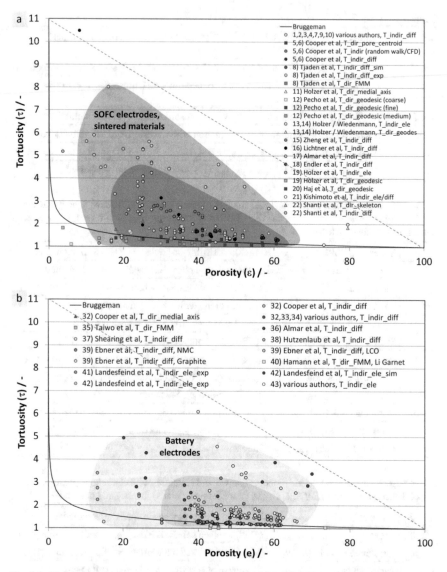

Fig. 3.1 Plots of tortuosity versus effective porosity for different materials and microstructures, as presented in Tables 3.1. **a** SOFC electrodes and porous ceramics from Table 3.1a, **b** Battery electrodes from Table 3.1c, **c** PEM GDL and fibrous materials from Table 3.1b, **d** Rocks and geological materials from Table 3.1d, **e** 2D granular materials from Table 3.1e, **f** 3D granular materials from Table 3.1f, **g** Stochastic models of networked structures and open foams from Table 3.1g, **h** For each material type a characteristic field is indicated with a specific color. The underlying source file for Fig. 3.1, with detailed information to 69 references, can be downloaded from the electronic appendix (Supplementary file 3.1)

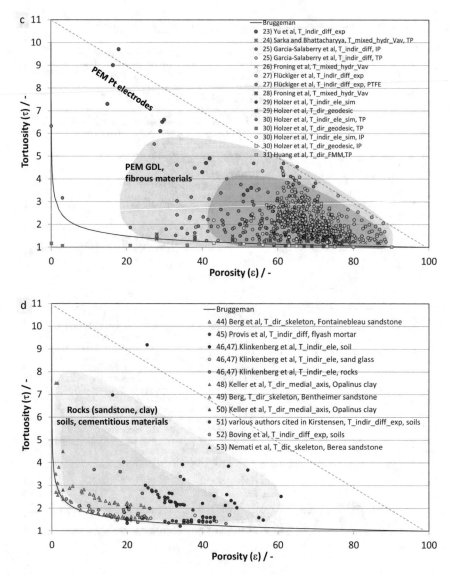

Fig. 3.1 (continued)

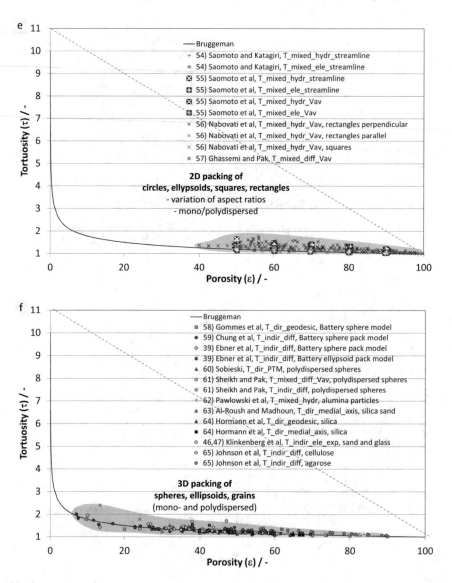

Fig. 3.1 (continued)

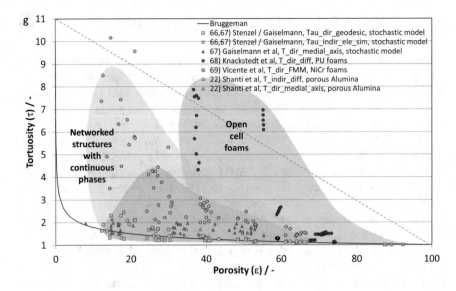

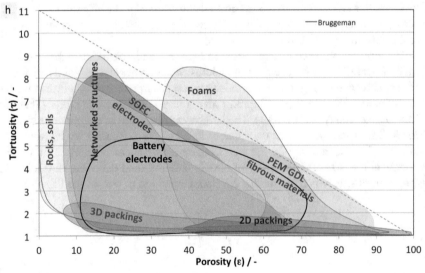

Fig. 3.1   (continued)

3.3   Empirical Data for Different Tortuosity Types

The same empirical data from Tables 3.1, which in Fig. 3.1 is grouped into different material types, is now replotted in Fig. 3.2, and thereby regrouped into different tortuosity types. The results in Fig. 3.2 illustrate a pronounced pattern, whereby the indirect tortuosities (Fig. 3.2a) scatter over a much larger range than the mixed (Fig. 3.2b) and geometric (Fig. 3.2c, d) tortuosity types.

In Fig. 3.2e geometric and mixed tortuosities are plotted together. But compared to Fig. 3.2a–d, they are plotted with a different scale on the y-axis. The characteristic field in red, representing geodesic and FMM tortuosities, shows the lowest tortuosity values. The measured values for geodesic and FMM tortuosities are often slightly below the Bruggeman trend line and only rarely they take values larger than 2. In contrast, the Bruggeman trend line typically defines the lower bound for mixed tortuosities (green characteristic field) and for those geometric tortuosities, which are derived from a skeleton (medial axis and PTM, blue characteristic field). For porosities below 0.3, the medial axis/PTM tortuosities often also reach values greater than 2. It must be mentioned here that for mixed tortuosities, the empirical data at low porosities is missing, but it is expected that the mixed tortuosity values also show a strong increase with decreasing porosity, in a similar way as it is observed, e.g., for medial axis and PTM tortuosities.

The comparison of empirical results in Figs. 3.1 and 3.2, indicate that the large scatter observed for most materials and microstructure types can be mainly attributed to data points containing indirect tortuosity types. A surprising insight is the fact that the different tortuosity types (Fig. 3.2) give a more characteristic pattern than the different material types (Fig. 3.1). The latter were discussed in the previous section. *This indicates that in general the value of tortuosity is more strongly influenced by the type of tortuosity than by the type of material or microstructure.* This finding strongly emphasizes the necessity of clearly defining (and choosing carefully) the type of tortuosity, which is used for the characterization of porous media.

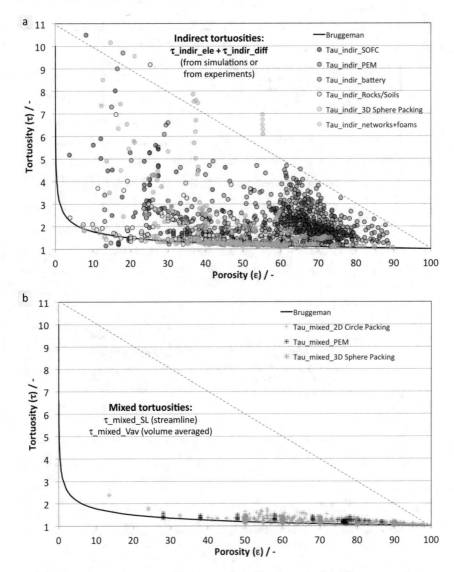

Fig. 3.2 Plots of tortuosity versus effective porosity with color codes for different tortuosity types (data from Tables 3.1). **a** Indirect electrical and diffusional tortuosities, **b** Mixed tortuosities, **c** Direct geometric tortuosities (medial axis, skeleton and PTM), **d** Direct geodesic tortuosities, **e** Comparison of mixed and direct tortuosity types (characteristic fields are highlighted with different colors). Note that in Fig. 3.2e, the y-axis has a different scale compared to Fig. 3.2a–d

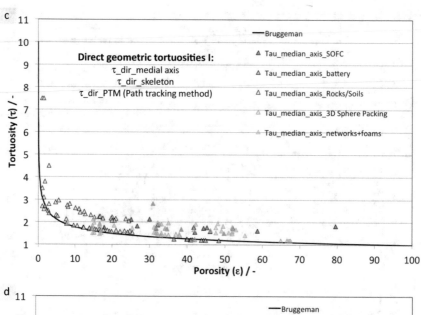

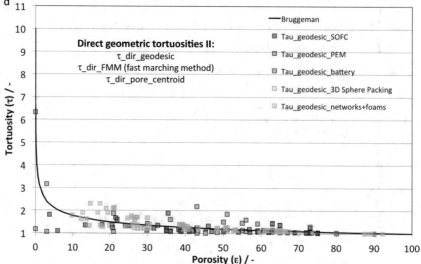

Fig. 3.2 (continued)

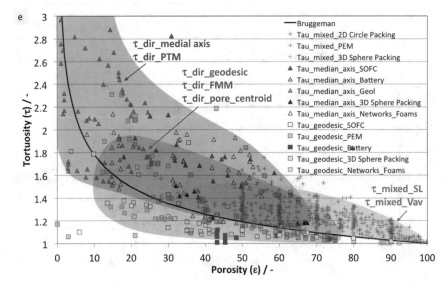

Fig. 3.2 (continued)

3.4 Direct Comparison of Tortuosity Types Based on Selected Data Sets

The data reviewed in the previous sections documents a kind of universal pattern with systematic differences between the various tortuosity types. In this section, we focus on examples that illustrate the difference between specific tortuosity types for a given material and microstructure. This investigation is based on selected data sets from Tables 3.1, where more than one tortuosity type is characterized for the same material or microstructure. For each example, we refer to the corresponding (Nr of data set) in Tables 3.1, which correspond to the [Ref Nr].

3.4.1 Example 1: Indirect Versus Direct Pore Centroid Tortuosity

τ_{indir_diff} versus $\tau_{dir\_pore\_centroid}$ from SOFC and battery electrodes.

Cooper et al. [5, 6] (dataset Nr 6a-e) used 5 different methods to extract tortuosity from an SOFC cathode material (LSCF). The data set includes *four indirect tortuosities* that were determined with different simulation approaches (6b: τ_{indir_therm} StarCCM+, mesh-based, 6c: τ_{indir_diff}, AvizoXlab, voxel-based, 6d: τ_{indir_diff}, TauFactor, voxel-based, 6e: τ_{indir_diff}, random walk algorithm, voxel-based). As shown in Fig. 3.3, all four approaches give very similar results for the

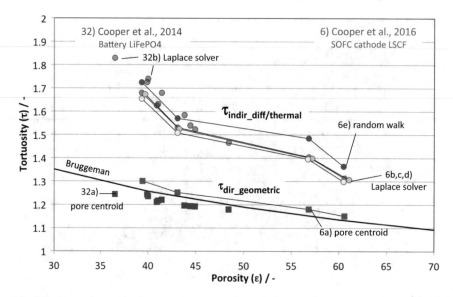

Fig. 3.3 Comparison of indirect (circles) versus direct tortuosities (squares) for SOFC cathodes (red symbols indicating values from Cooper et al. [5, 6]) and for battery electrodes (blue symbols indicating values from Cooper et al. [32]). Both materials show the same pattern, where the values for indirect tortuosity ($\tau_{indir\_diff\_sim}$) are consistently higher than those for geometric tortuosity ($\tau_{dir\_pore\_centroid}$). Note that the values for geometric tortuosity are very close to the Bruggeman trend line

indirect tortuosities (red circles). In comparison, the direct geometric tortuosity (6a: $\tau_{dir\_pore\_centroid}$) for the same sample is significantly lower (red squares).

Cooper et al. [32] (dataset Nr 32a-b, see Fig. 3.3, blue symbols) also presented a comparison of direct geometric tortuosities (32a: $\tau_{dir\_pore\_centroid}$) with indirect diffusional tortuosities (32b: τ_{indir_therm}) for a Li-ion battery electrode (LiFePO$_4$). The overall pattern and even the specific tortuosity-porosity values are very similar when comparing the battery electrode (blue) with the SOFC electrode (red). For both materials the values for indirect tortuosity ($\tau_{indir\_diff/therm}$) are consistently higher than those for geometric tortuosity ($\tau_{dir\_pore\_centroid}$). From this data, Cooper et al. [5] deduced the following relationship between indirect and direct tortuosities:

$$\tau_{dir\_pore\_centroid} = 0.5 \ln \tau_{diff\_indir} + 1 \qquad (3.1)$$

3.4.2 Example 2: Indirect Versus Direct Medial Axis Tortuosity

$\tau_{indir\_ele\_exp}$ versus $\tau_{dir\_medial\_axis}$ from porous ceramic membranes.

Wiedenmann et al. [14] (Dataset Nr. 14 in Table 3.1) presented a comparison of direct medial axis tortuosities (14a: $\tau_{dir\_medial\_axis}$) versus indirect electrical tortuosities (14b: τ_{indir_ele}, from EIS experiments) for two different separation membranes in alkaline electrolysis cells, consisting of sintered olivine and wollastonite. As shown in Fig. 3.4 the microstructures vary from fine grained and dense ($\varepsilon = 0.27$) to coarse-grained and open porous ($\varepsilon = 0.80$). Despite the large variation of porosity, the values for medial axis tortuosity (filled squares) are all in a very narrow range (1.62–1.84). The 3D visualizations in Fig. 3.4 [14] document nicely, that the basic geometry of all pore networks remains very similar for all samples, except for the coarseness, which increases with porosity (i.e., scaling). When changing porosity, the obtained values for geometric tortuosity thus remain almost constant (1.73 ± 0.11).

In contrast, the indirect tortuosities (open circles) increase significantly with decreasing porosity from 1.4 to 2.2 for olivine (blue) and from 1.6 to 2.4 for wollastonite (red). Wiedenmann et al. [14] and Holzer et al. [13] documented that the effective properties in these samples scale with constrictivity (i.e., bottlenecks), but not with geometric tortuosity. Therefore, they concluded that the variation of indirect

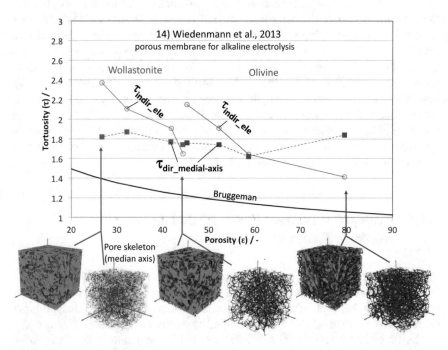

Fig. 3.4 Comparison of indirect ($\tau_{indir\_ele\_exp}$) versus direct tortuosities ($\tau_{dir\_medial\_axis}$) for porous membranes consisting of wollastonite (red) and olivine (blue) from Wiedenmann et al. [14]. The geometric tortuosities (filled squares) vary in a narrow range (1.6–1.8), whereas the indirect tortuosities (open circles) show significant variation (1.4–2.4). The variation of effective properties and associated indirect tortuosities is attributed to variations of the size of characteristic bottlenecks and corresponding constrictivities

tortuosities in this example rather represents the resistive effects arising from variations in the size of characteristic bottlenecks, rather than effects from path length
variations.

3.4.3   Example 3: Indirect Versus Direct Geodesic Tortuosity

$\tau_{indir\_ele/therm\_sim}$ versus $\tau_{indir\_diff\_exp}$ versus $\tau_{dir\_geodesic/FMM}$ from porous Zr-oxide and
PEM GDL.

Figure 3.5 represents a comparison of direct geometric tortuosities (geodesic/
FMM) with indirect tortuosities (diffusive/electric/thermal—from simulation and
experiment) based on data from Holzer et al. [19], Tjaden et al. [8] and Holzer et al.
[29, 30]. As shown in Fig. 3.5, all four studies show that the values for geometric
tortuosities ($\tau_{dir\_geodesic}$ and τ_{dir_FMM}, closed squares) are systematically lower than
the results for indirect tortuosity (open circles and crosses). The samples of the
three studies cover a wide range of effective porosities from 4 to 76 vol-%. In addition, the involved microstructures in sintered ceramics (Zr-oxide, YSZ) are very
different from those in fibrous PEM GDL. Despite this obvious microstructural variation, all samples show the same consistent pattern. The direct geometric tortuosities
(geodesic/FMM) vary hardly and are always below the Bruggeman trend line. Only
for very low porosities ($\varepsilon < 0.2$) the geodesic tortuosities start to increase moderately.
Apparently, the geodesic path lengths are not very sensitive to variation of the pore
volume fraction. In contrast the indirect tortuosities (open symbols and crosses) are
always significantly higher than the Bruggeman trend line and they also show much
stronger variation. For each series of material/microstructure, a trend of increasing
indirect tortuosities is observed when lowering the porosity.

3.4.4   Example 4: Indirect Versus Medial Axis Versus
Geodesic Tortuosity

$\tau_{indir\_ele\_sim}$ versus $\tau_{dir\_medial\_axis}$ versus $\tau_{dir\_geodesic}$ from stochastic 3D structures.

Gaiselmann et al. [67] and Stenzel et al. [66] performed in-depth investigations on
the relationship between microstructure characteristics and effective transport properties based on virtual 3D structures generated by stochastic modelling. The so-called
stochastic spatial graph model (SSGM) provides 3D structures with a connected
transporting phase even at low volume fractions. A large number of microstructures covering a wide range of microstructure characteristics (i.e., volume fractions,
phase size distributions, constrictivity and path lengths) were created and analysed.
Three different tortuosities (medial axis, geodesic, and indirect tortuosities) can be
compared based on datasets Nr. 66 and Nr. 67. It should be noted, that in these studies,
electric conduction in the solid phase and associated solid phase microstructure was

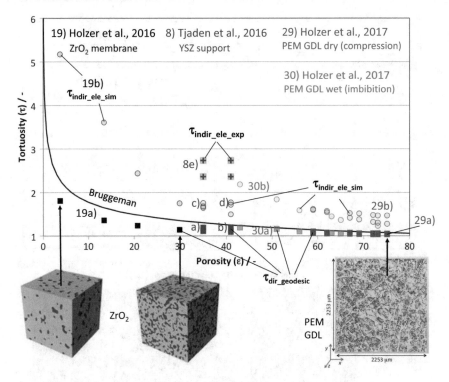

Fig. 3.5 Comparison of indirect ($\tau_{indir\_ele\_exp/sim}$) versus direct geometric tortuosities ($\tau_{dir\_geodesic}$, τ_{dir_FMM}) for various porous materials from Holzer et al. [19] (black symbols), Tjaden et al. [8] (blue symbols), and Holzer et al. [29, 30] (red and green symbols). The direct geometric tortuosities (closed symbols) vary only in a narrow range below the Bruggeman trend line. In contrast, for the same samples, the indirect tortuosities (open symbols) show much higher values

investigated. However, the effect of microstructure (e.g., tortuosity) on transport in solid phases is basically the same as in the pore phase.

The results in Fig. 3.6 document that the indirect tortuosities ($\tau_{indir\_ele\_sim}$) are consistently higher than the direct geometric tortuosities. Figure 3.6 also reveals slight differences between direct medial axis and direct geodesic tortuosities. The medial axis tortuosity ($\tau_{dir\_medial\_axis}$, red symbols, Nr 67) is slightly higher than the Bruggeman trend line. In contrast, the geodesic tortuosity ($\tau_{dir\_geodesic}$, black, Nr 66b) is usually below the Bruggeman trend line. However, for small volume fractions of the transporting phase (i.e., $\varepsilon < 20$ vol-%), both tortuosities (i.e., geodesic, and medial axis) show similar values.

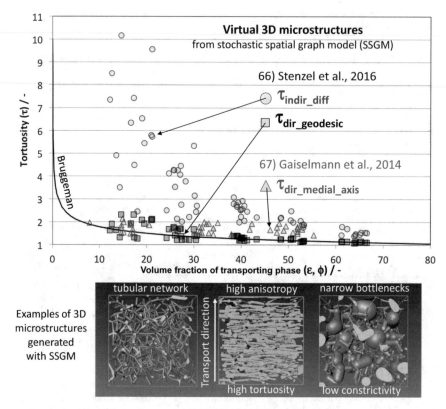

Fig. 3.6 Comparison of indirect ($\tau_{indir\_ele\_sim}$) versus direct geometric tortuosities ($\tau_{dir\_geodesic}$, $\tau_{dir\_medial\_axis}$) for virtual 3D structures created with a spatial stochastic graph model (SSGM) from Gaiselmann et al. [67] and Stenzel et al. [66]. The direct geometric tortuosities (black squares, red triangles) vary in a narrow range, which is close to the Bruggeman trend line. For the same samples, the indirect tortuosities (blue open circles) show much higher values and a stronger variation

3.4.5 Example 5: Direct Medial Axis Versus Direct Geodesic Tortuosity

$\tau_{dir\_medial\_axis}$ versus $\tau_{dir\_geodesic}$ from SOFC anodes and silica monoliths.

The relationship between two direct geometric tortuosities (i.e., geodesic, and medial axis tortuosities) is also investigated by Holzer et al. [11] and Pecho et al. [12] for SOFC anodes and by Hormann et al. [64] for silica monoliths. Figure 3.7 clearly shows that geodesic tortuosities (squares) are systematically lower than the medial axis tortuosities (triangles), even though they are measured for the same samples. Furthermore, the Bruggeman trend line can be roughly considered as the boundary between the characteristic fields for these two geometric tortuosity types. These findings are compatible with the results from Gaiselmann et al. [67] and Stenzel et al. [66] in Example 4 (Fig. 3.6).

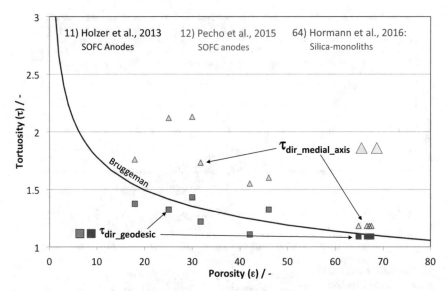

Fig. 3.7 Comparison of geodesic ($\tau_{dir\_geodesic}$) versus medial axis tortuosities ($\tau_{dir\_medial\_axis}$) for SOFC anodes (Holzer et al. [11] and Pecho et al. [12]) and for silica monoliths (Hormann et al. [64]). The geodesic tortuosity ($\tau_{dir\_geodesic}$) typically shows values close to or below the Bruggeman trend line. In contrast, the medial axis tortuosities ($\tau_{dir\_medial\_axis}$) show consistently higher values

3.4.6   Example 6: Mixed Streamline Versus Mixed Volume Averaged Tortuosity

$\tau_{mixed\_ele/hydr\_streamline}$ versus $\tau_{mixed\_ele/hydr\_Vav}$ (mixed tortuosity types) for simulated particle packing (2D and 3D).

This example uses datasets with mixed tortuosity types, including the streamline tortuosity ($\tau_{mixed\_phys\_streamline}$) and the volume-averaged tortuosity ($\tau_{mixed\_phys\_Vav}$) for flow, conduction, and diffusion.

Saomoto et al. [55] simulated hydraulic flow and electrical conduction in simple 2D structures consisting of mono-sized circles and/or ellipsoids. Both mixed tortuosities (i.e., streamline and area-averaged) are extracted from the electric and hydraulic flow fields and their values are plotted in Fig. 3.8. Surprisingly the results for both mixed tortuosity types are almost identical (if considering results for one specific type of transport). For example, the *electric* streamline tortuosity (red square with crosses) is nearly identical to the *electric* volume-averaged tortuosity (blue square with crosses) (i.e., $\tau_{mixed\_ele\_streamline} \cong \tau_{mixed\_ele\_Vav}$). The same holds for hydraulic tortuosities (i.e., $\tau_{mixed\_hydr\_streamline} \cong \tau_{mixed\_hydr\_Vav}$). However, a significant difference is observed between electric and hydraulic tortuosities. The characteristic field for mixed *electric* tortuosities (i.e., $\tau_{mixed\_ele\_streamline}$, $\tau_{mixed\_ele\_Vav}$), which is highlighted in red color, is lower than the characteristic field for mixed *hydraulic* tortuosities (i.e., $\tau_{mixed\_hydr\_streamline}$, $\tau_{mixed\_hydr\_Vav}$), which is highlighted in blue. The boundary between these characteristic fields is roughly identical with the Bruggeman trend line.

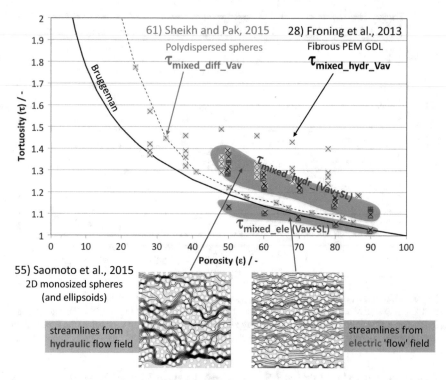

Fig. 3.8 Comparison of different mixed tortuosities ($\tau_{mixed\_ele/diff/hydr\_streamline}$, $\tau_{mixed\_ele/diff/hydr\_Vav}$). In principle, the datasets for mono-sized spheres (and ellipsoids) from Saomoto et al. [55] (dataset 55, 2D) and from Sheikh and Pak [61] (dataset 61, 3D) show hardly a difference between streamline and volume averaged tortuosities. However, a systematic difference is caused by the underlying transport mechanism, in the sense that mixed hydraulic tortuosities (characteristic field in blue, and yellow crosses) are higher than the corresponding mixed electric tortuosities (characteristic field in red)

Sheikh and Pak [61] (green crosses) reported *diffusive* volume-averaged tortuosities ($\tau_{mixed\_diff\_Vav}$) from 3D poly-dispersed spheres, which are close to the values of mixed *electric* tortuosities in Saomoto et al. [55], but lower than the mixed *hydraulic* tortuosities in [55]. This finding is compatible with the general expectation that diffusive and electrical transport properties and associated tortuosities are almost identical to each other and that the hydraulic tortuosities are generally somewhat higher, see e.g., Clennell [71].

Finally, results of volume averaged hydraulic tortuosities ($\tau_{mixed\_hydr\_Vav}$) for gas diffusion layers (GDL) in PEM fuel cells are presented by Froning et al. [28] (crosses highlighted in yellow). Note that the microstructures of fibrous GDL considered by Froning et al. [28] are significantly different from those of poly-dispersed sphere packing in Sheikh and Pack [61]. Nevertheless, the characteristic values of $\tau_{mixed\_hydr\_Vav}$ for these two materials (GDL in [28] and packed spheres in [61]) are rather similar. This shows again that the tortuosity values are usually more strongly depending on the type of tortuosity and less strongly on the type of material and/or microstructure.

Overall, the results presented in this example show that the values of mixed tortuosities are relatively low (i.e., close to the Bruggeman trend line). Values larger than 2 are rarely observed and can only be expected for structures with either low porosities ($\varepsilon < 0.2$) or with strong anisotropy effects. Similar as observed previously for the various geometric tortuosities, also the mixed tortuosities show a relatively small scatter, and the values are usually in the range between 1 and 2 (i.e., roughly compatible with Carman's estimation of $\sqrt{2}$). This is in remarkable contrast to the indirect tortuosities (not analysed in this example), which scatter over a much larger range and often take values much larger than 2.

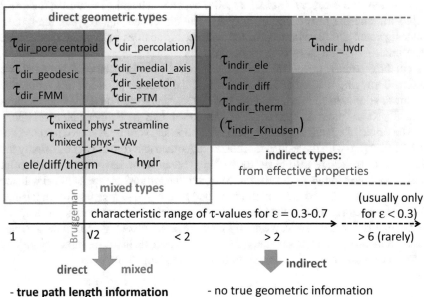

Fig. 3.9 The review of tortuosity-porosity data from literature reveals a universal pattern that is characterized by a consistent relative order among the different tortuosity types. This pattern is observed for many different materials and microstructures. Geometric and mixed tortuosity types typically show relatively low values close to the Bruggeman trend line. In contrast, the indirect tortuosities show higher values with a larger variability

3.5   Relative Order of Tortuosity Types

3.5.1   Summary of Empirical Data: Global Pattern of Tortuosity Types

The empirical data from literature reveals a *consistent pattern among the tortuosity types*, in the sense that certain tortuosity types give consistently higher values than others. This relative order of tortuosity types, which is schematically illustrated in Fig. 3.9, is valid for a wide range of materials with very different microstructures.

Basically, the *indirect tortuosities* scatter over a much wider range than the direct geometric and the mixed tortuosity types. The direct and mixed types rarely take values greater than 2, whereas for the indirect tortuosities much larger values are measured,—in some cases even higher than 20.

For the *geometric tortuosities*, two subgroups can be distinguished. For medial axis and path tracking method (PTM) tortuosities, the Bruggeman trend line represents the lower bound. In contrast, geodesic and fast marching method (FMM) tortuosities typically show values that are equal or even lower than the Bruggeman trend line.

The values of *mixed tortuosity* types (streamline and volume averaged tortuosities) roughly overlap with the values for the direct geometric tortuosities (i.e., usually the mixed tortuosities are also close to the Bruggeman trend line—see e.g., Fu et al. [72]). Example 6 indicates that the mixed streamline tortuosities are identical with the mixed volume-averaged tortuosities, provided that the same transport mechanism is considered (see e.g., Saomoto et al. [55]). However, the mixed electrical, diffusional, or thermal tortuosities are consistently lower than the mixed hydraulic tortuosities. The Bruggeman trend line separates the two characteristic fields for mixed electrical/ diffusional and for mixed hydraulic tortuosities (see Fig. 3.8).

3.5.2   Interpretation of Different Tortuosity Categories

3.5.2.1   Direct and Mixed Tortuosities

The *direct and mixed tortuosities* are based on geometric analyses and therefore they can be considered as true or realistic measures for the path lengths through the porous medium under investigation. Consequently, in order to predict the impact of microstructure on effective transport properties, it is not sufficient to merely consider the geometric or mixed tortuosities, since other morphological effects (e.g., bottlenecks/constrictivity) in addition to path length variation also have an influence on the effective transport properties. Predictions of effective transport properties based on distinct estimations of the path length effect (tortuosity), bottleneck effect (constrictivity) and viscous drag at pore walls (hydraulic radius) are extensively discussed in Chap. 5. Using tortuosity types that give a realistic estimation of the true path length is the key to understanding and differentiating different microstructure effects that limit the transport in porous media.

3.5.2.2   Indirect Tortuosities

The *indirect tortuosities* are derived from effective (or relative) properties that are known from experiment or simulation. The indirect tortuosities can thus be considered as a measure for the bulk microstructure resistance against transport. The large values and the large variability observed for indirect tortuosities are due to the fact, that they capture all different kinds of microstructure limitations, including resistive effects from narrow bottlenecks. Indirect tortuosities are thus not a realistic measure for the length of transport paths since they tend to overestimate the effect of pure path lengths significantly. Indirect tortuosities can also be interpreted as fudge factor that describes the ratio of relative transport property over porosity (i.e., $\sqrt{(\sigma_{rel}/\varepsilon)}$ or $\sqrt{(D_{rel}/\varepsilon)}$). For viscous flow, the indirect hydraulic tortuosity is rarely calculated, since viscous drag expressed by hydraulic radius makes calculation more complicated (see discussion in Chap. 2).

3.6   Tortuosity–Porosity Relationships in Literature

3.6.1   Mathematical Expressions for τ–ε Relationships and Their Limitations

Numerous mathematical expressions describing tortuosity–porosity (τ–ε) relationships can be found in literature. The different τ–ε relationships are reviewed by Shen and Chen [73], Ghanbarian [74], Tjaden et al. [75] and Idris et al. [76]. Table 3.2 represents a selection of mathematical $\tau - \varepsilon$ relationships from literature.

Note that very different mathematical expressions are proposed—usually logarithmic and power-law functions. As shown in Fig. 3.10a, the resulting τ–ε curves diverge greatly from each other. It may be argued that the observed variety results from the fact, that these τ–ε relationships are derived for different material types and different microstructures (e.g., battery electrodes, clays). However, most of these relationships are derived for packed spheres.

The observed chaotic picture of mathematical expressions (Fig. 3.10a) contrasts with the empirical data, which shows a clear pattern when plotted separately for different tortuosity types, as summarized in Fig. 3.9. Moreover, the empirical data typically results in characteristic τ–ε fields (as shown in Fig. 3.2 for the different tortuosity types), but usually it does not result in clearly defined τ–ε curves. This is particularly not the case when plotted for specific material types (see Fig. 3.1) and thereby not distinguishing the involved tortuosity types.

Table 3.2 Mathematical tortuosity—porosity (τ–ε) relationships from literature

Nr.	τ type	τ^2 (or τ!) = $f(\varepsilon)$	Material type or microstructure type	x = ..	Reference(s)
1	Geom. model	$\tau_{geometric} = \frac{1}{2}\left[1 + \frac{1}{2}\sqrt{1-\varepsilon} + \frac{\sqrt{(1-\sqrt{1-\varepsilon})^2+(1-\varepsilon)/4}}{1-\sqrt{1-\varepsilon}}\right]$	Geometric 2D model of square particles		Yu and Li [81]
3, 2	Geom. model	$\tau_{geometric} = 1.23\dfrac{(1-\varepsilon)^{4/3}}{x^2\varepsilon}$	Packed particles → Packed spheres →	0.75 1	Lanfrey et al. [82]
4	Geom. model	$\tau_{geometric} = \left(\dfrac{19}{18}\right)^{\ln(\varepsilon)/\ln(8/9)}$	Pore fractal model		Li and Yu [83]
5, 6, 7	Hydr	$\tau_{hydr}^2 = 1 - x\ln(\varepsilon)$	Spheres + fibers → Plates + flakes → High aspect ratio particles →	0.5 1 3	Pech and Renaud, cited in Comiti and Renaud [84]
8	Hydr	$\tau_{hydr}^2 = \varepsilon^{-x}$	Binary mixture of spheres	0.4	Mota et al. [85]
9	Hydr	$\tau_{hydr}^2 = \dfrac{\varepsilon}{1-(1-\varepsilon)^{2/3}}$	Isotropic granular material		Du Plessis et al. [86]
10, 11	Hydr	$\tau_{hydr}^2 = \sqrt{\dfrac{2\varepsilon}{3\left[1-x(1-\varepsilon)^{2/3}\right]} + \dfrac{1}{3}}$	Monosized spheres: Cubic packing → Tetrahedral packing →	1.209 1.108	Ahmadi [87]
12	Hydr	$\tau_{mix\_hyd\_Vav} = 1 + 0.8(1-\varepsilon)$	Mono-sized solid rectangles in 2D		Koponen et al. [88]
13	Electr	$\tau_{ele}^2 = 1 + 0.5(1-\varepsilon)$ $\tau_{ele}^2 = \dfrac{(3-\varepsilon)}{2}$	Dilute suspension of spheres		Maxwell, 1873 [89]
14, 15	Electr	$\tau_{ele}^2 = \varepsilon^{1-x}$	Cementation Exponents: For rocks → For sediments →	1.1 2	Archie, 1942 [79]

(continued)

Table 3.2 (continued)

Nr.	τ type	$\tau^2 (or \tau) = f(\varepsilon)$	Material type or microstructure type	x = ..	Reference(s)
16	Electr	$\tau_{ele}^2 = \varepsilon^{1-x}$, **identical with Bruggeman (see Nr 27)**	**Packing of poly-dispersed spheres**	**1.5**	**Archie, 1942 [79]**
17, 18, 19, 20, 21, 22	Electr	$\tau_{ele}^2 = 1 - x\ln(\varepsilon)$	Spherical particles → Monosized spheres → Cubic particles → Cylinders → Overlapping spheres → Monosized spheres →	0.41 0.49 0.63 1.00 0.5 0.5	Pech and Renaud, cited in Comiti and Renaud [84]
23	Diff	$\tau_{diff}^2 = \varepsilon^{-x}$	Mono-sized spheres	0.4	Mota et al. [85], Delgado [90]
24	Diff	$\tau_{diff}^2 = \frac{\varepsilon}{1-(1-\varepsilon)^{1/3}}$	Catalyst		Beeckman [91]
25	Diff	$\tau_{diff}^2 = \varepsilon + x(1 - \varepsilon)$	Sand, silt, sediments	2	Iversen and Jørgensen [92]
26	Non spec	$\tau^2 = 2 - \varepsilon$			Petersen [93]
27	Diff	$\tau_{diff}^2 = \varepsilon^{-1/2} = \varepsilon^{1-x}$ **for spheres identical with Archie (See Nr 16)**	Polydisperse granular media: Spheres → Cylinders →	**1.5** 2	**Bruggeman 1935 [94]** see also Tjaden et al. [95]
28	Electr	$\tau_{diff}^2 = x_1 \varepsilon^{1-x_2}$	Battery electrodes and separators	x_1: 0.1–2.6 x_2:1.27–5.2	Thorat [96], cit. in Tjaden [75]
29	Electr	$\tau_{ele}^2 = \varepsilon^{-x}$	Battery electrode	0.5–2	Ebner [39]
30	Diff	$\tau_{diff}^2 = \varepsilon^{-1/3}$	Monodisperse granular media		van Brakel and Heertjes [97]
31	Hydr	$\tau_{diff}^2 = 1 - \ln(\varepsilon)$	Overlapping cylinders		Tomadakis and Sotirchos [98]
32	Diff	$\tau_{diff}^2 = 1 - \frac{\ln(\varepsilon)}{2}$	Overlapping spheres		Weissberg [99]
33	Diff	$\tau_{diff} = 1 - x\ln(\varepsilon)$	Clays dry and hygroscopic	0.357 0.503	Sun et al. [100]

(continued)

Table 3.2 (continued)

Nr.	τ type	$\tau^2 (or\ \tau!) = f(\varepsilon)$	Material type or microstructure type	x = ..	Reference(s)
34	Hydr	$\tau^2_{hydr} = \left(\dfrac{2-\varepsilon}{\varepsilon}\right)^2$	Cation exchange resin		Mackie [101]
35	Diff	$\tau_{diff} = \left[1 - x(1-\varepsilon)^{-1}\right]$	Granular media with: spheres → Cubes → Large and → Small parallelepipeds →	0.6 0.73 1.07 1.21	Pisani [102]
36	Hydr	$\tau^2_{mixed\_hydr\_Vav} = x_1 - x_2 \ln(\varepsilon^2)$	Monosized spheres	x_1: 1.1842 x_2: 0.6579	Jin et al. [103]
37	Hydr	$\tau^2_{mixed\_hydr\_Vav} = x_1 - x_2 \ln(\varepsilon^2)$	Monosized spheres	x_1: 0.9463 x_2: 0.7173	Duda [104], cit. in Jin [103]
38	Hydr	$\tau_{mixed\_hydr\_Vav} = x_1(1 - \varepsilon) + x_2$	Monosized spheres	x_1: 0.8002 x_2: 1.0454	Jin et al. [103]
39	Hydr	$\tau_{mixed\_hydr\_Vav} = x_1(1 - \varepsilon) + x_2$	Monosized spheres	x_1: 0.9119 x_2: 0.9340	Duda [104], cit. in Jin [103]
40	Hydr	$\tau_{mixed\_hydr\_Vav} = x_1 - x_2 \ln(\varepsilon)$	Monosized spheres	x_1: 1.1133 x_2: 0.4845	Jin et al. [103]
41	Hydr	$\tau_{mixed\_hydr\_Vav} = x_1 - x_2 \ln(\varepsilon)$	Monosized spheres	x_1: 1.0104 x_2: 0.5541	Duda [104], cit. in Jin [103]
42	Hydr	$\tau_{mixed\_hydr\_Vav} = 1 - x_1 \ln(\varepsilon)$	2D monosized spheres/cubes	0.5/0.541	Saomoto and Katagiri [54]
43	Ele	$\tau_{mixed\_ele\_Vav} = 1 - x_1 \ln(\varepsilon)$	2D monosized spheres/cubes	0.2/0.19	Saomoto and Katagiri [54]

It must be emphasized that in most references the tortuosity type is not defined clearly. Information on 'τ type' is thus often interpreted from the context of a given paper, where possible. 'Material type' describes the microstructure for which the τ–ε relationship was derived. Note that depending on the authors, the lefthand side of the equation is sometimes written as tortuosity (τ) and sometimes as tortuosity factor (τ^2)

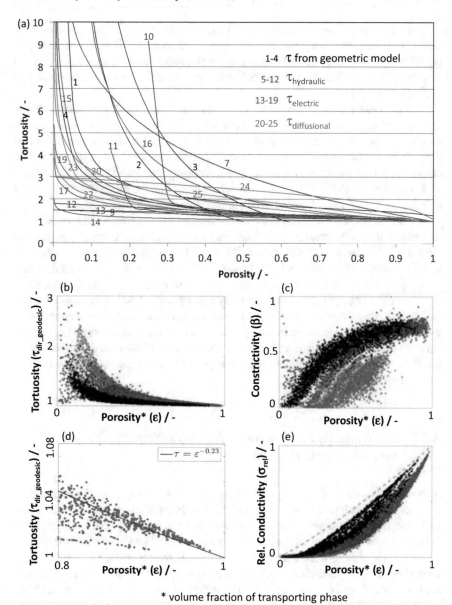

* volume fraction of transporting phase

Fig. 3.10 a Mathematical tortuosity–porosity (τ–ε) relationships from literature. The numbers refer to the equations given in Table 3.2. Colors indicate which type of 'physics-based' tortuosity is considered. It must be emphasized that in the source papers it is usually not indicated, whether these are indirect, mixed, or direct tortuosities, **b, c, d** and **e** plots of tortuosity ($\tau_{dir\_geodesic}$), constrictivity (β), and relative electric conductivity (σ_{rel}) versus porosity (ε, on the x-axis). Red, blue, and black data points represent three distinct types of 3D microstructures, which are generated with different stochastic models. More than 2000 different 3D structures are investigated, which cover a wide range of microstructure characteristics (Figs. 3.10b, c, e are taken from Neumann et al. [77], and Fig. 3.10d is from Stenzel et al. [78])

The large scatter for the mathematical expressions in Fig. 3.10a thus illustrates that there is no consensus how tortuosity varies with porosity. In fact, this finding is questioning the underlying assumption that variations of tortuosity and porosity are strictly related, so that the τ–ε relationship could be described with one universal mathematical formula.

The behaviour of microstructure characteristics with varying porosity can be investigated in a more general way, when 3D image analysis is applied to a large number of 3D microstructure models. Results from such studies are shown in Fig. 3.10b–e (taken from Neumann et al. [77] and Stenzel et al. [78]). It must be emphasized that these investigations are using only one specific tortuosity type, which is geodesic tortuosity. These data sets indicate that the microstructure characteristics (ε, β, $\tau_{dir\_geodesic}$) can vary in a large range within the theoretically possible constellations, except for high porosities, where tortuosity asymptotically tends to 1 (see Fig. 3.10d). For small porosities the variation of geodesic tortuosity is particularly large, but also at very high porosities ($\varepsilon > 0.8$) there is still significant variation in the tortuosity values (Fig. 3.10d). A similar behaviour is expected for all the other direct geometric and mixed tortuosities. These examples thus indicate that there exists no clear correlation between porosity and tortuosity (and associated effective path length, respectively), which could be described with a universal mathematical τ–ε expression. As shown in Fig. Fig. 3.10e, there exists also no strict correlation between relative conductivity and porosity. Hence, it is not justified to expect a strict correlation between indirect tortuosity and porosity, since the first one is derived from relative conductivity.

3.6.2 Mathematical Expressions for τ–ε Relationships and Their Justification

Nevertheless, mathematical expressions for tortuosity–porosity (τ–ε) relationships may have their justification in context with specific cases of controlled microstructure variation. Such a case was described by Archie [79], who presented experimental data for porous sediments saturated with an electrolyte. For the special case, where all samples originate from the same sedimentary unit, the experimental results show a correlation between electric resistance and porosity. Archie's law (see Eqs. 2.22 and 2.23: $F_R = \varepsilon^{-m} = 1/\sigma_{rel}$) describes this correlation using a so-called cementation factor (m) as exponent. Wyllie and Rose [80] redefined the relationship between electric resistance (formation factor, F_R, respectively) and porosity by introducing the so-called structural factor (Eq. 2.24: $F_R = X_{ele}/\varepsilon = 1/\sigma_{rel}$). Thereby, X_{ele} can be interpreted as being equivalent with Carman's tortuosity factor ($X_{ele} = T = \tau_{ele}{}^2$). This leads to Eq. 2.25 (i.e., $\tau_{indir\_ele}{}^2 = \varepsilon/\sigma_{rel}$), which is widely used to compute indirect tortuosity. Thereby, indirect tortuosity can be considered as some kind of proportionality between porosity and effective properties (e.g., transport resistance or conductivity). For the special case where effective or relative properties (F_R, σ_{rel})

correlate with porosity, it follows that there must be also a strict correlation with indirect tortuosity, such that Archie's law can be reformulated as follows:

$$\tau_{indir\_ele}^2 = \varepsilon^{1-m} \tag{3.2}$$

It must be emphasized that Archie's cementation factor (m > 1) is a fit parameter, which is valid only for a special case of microstructure variation. The cementation factor takes a specific value for a series of rocks, which all have undergone the same diagenetic process (i.e., solidification process transforming loose sediment into a solid rock). The common history of these sediments in the same geo-environment led to a characteristic variation of the microstructure, so that sizes of pores and bottlenecks, pore connectivity and transport path lengths vary in a characteristic way with porosity. Hence, in this case the correlation of microstructure with porosity is controlled by the diagenetic conditions and by the associated cementation process. Due to this controlled correlation, it is possible to find a suitable mathematical expression for the $\tau - \varepsilon$ relationship of rocks in a single sedimentary unit. Thereby the indirect tortuosity represents the lumped sum of all microstructure effects. With respect to Archie's law, it must be realized that the fitting of the cementation factor (m) has to be repeated if the rocks originate from a different sedimentary unit, because then these rocks were exposed to different diagenetic conditions, and therefore they are characterized by different $\tau - \varepsilon$ relationships.

Such special cases are also known from materials engineering when microstructure variations are performed in a controlled way. An example is described by Holzer et al. [19], where sintering temperatures are changed systematically, but all the other parameters (e.g., composition and sinter time) for the fabrication of porous ceramic membranes are kept constant. This results in a systematic variation of microstructure characteristics ($\tau_{dir\_geodesic}$, β) and effective properties (σ_{eff}, σ_{rel}) with porosity (ε). Hence, for such controlled microstructure variation a suitable mathematical expression can be found for the observed $\tau - \varepsilon$ relationship. However, this mathematical expression may no longer be valid, when one of the other fabrication parameters (e.g., composition or sinter time, grinding and particle size distribution of ceramic powders) is changed.

The Bruggeman equation is a special case of Archie's $\tau - \varepsilon$ relationship (see Eq. 3.2), where the 'cementation factor' (m) is fitted for a special microstructure type. For example, it was found that m = 1.5 for packed spheres (and m = 2 for cylindrical particles), which leads to

$$\tau_{indir\_ele}^2 = \varepsilon^{-0.5}. \tag{3.3}$$

In battery research, a modified version of the Bruggeman equation, given by

$$\tau_{indir\_ele}^2 = \gamma \varepsilon^{1-\alpha} \tag{3.4}$$

is nowadays often used (see Thorat et al. [96]), where γ is an additional fit parameter. Various authors [96, 105–110] presented experimental and numerical fits of Eq. 3.4

(e.g., Nrs 27–30 in Table 3.2) for different battery electrodes, which was critically reviewed by Tjaden et al. [75]. Thereby it is well illustrated that the variations of γ (0.1–2.6) and α (1.27–5.2) are very large and the resulting $\tau - \varepsilon$ curves differ significantly from each other, depending on the type of battery material. In addition, Tjaden et al. [75] also report examples from literature, which show that for many battery materials the Bruggeman equation and its modifications are simply not applicable. Tjaden et al. [75] thus concluded that tortuosity-porosity relationships such as the Bruggeman equation are only applicable to microstructures 'which are similar to the microstructure used to derive the respective relationship' (i.e., for special cases).

These selected examples related to Archie's and Bruggeman's equations illustrate that $\tau - \varepsilon$ relationships should not be mistaken as universal laws. In general, when varying materials microstructures are considered, the different microstructure characteristics (ε, β, τ, r_h, r_{min}, r_{max}) can vary independently from each other,—at least to some degree. Therefore, empirical data shows significant scatter of $\tau - \varepsilon$ datapoints for different material types (Fig. 3.1) and also for different tortuosity types (Fig. 3.2). A similar scatter of datapoints is also observed for other morphological characteristics (e.g., constrictivity) and effective properties (see Fig. 3.10b–e). The empirical data illustrate that, in general, there exists no simple and clear correlation between porosity and the other relevant microstructure characteristics (tortuosity, constrictivity, hydraulic radius).

Hence, we conclude that mathematical formulas for tortuosity-porosity relationships are valid only when the following special conditions are fulfilled:

(a) $\tau - \varepsilon$ relationship is defined for a specific type of tortuosity.

A suitable classification scheme and associated nomenclature are given in Sect. 2.6 (see Fig. 2.8). The empirical data show that the scatter of data points is generally much smaller for direct geometric and mixed tortuosity types (Fig. 3.2b–e), compared to the indirect tortuosities (Fig. 3.2a).

(b) $\tau - \varepsilon$ relationship is defined for a specific type of material and microstructure.

Thereby, the materials under consideration fulfil a systematic microstructure variation (e.g., rocks from the same sedimentary unit, which all had similar conditions during sedimentation and diagenesis). For simple microstructures such as packed spheres it is more probable that microstructure variation (i.e., densification) results in a systematic correlation between tortuosity and porosity (Fig. 3.1e, f), compared to more complex microstructures (e.g., SOFC electrodes, fibrous materials, foams), which tend to show a larger scatter (Figs. 3.1g, h) of τ-ε couples.

3.7 Summary

Nowadays, there are numerous methods available for the characterization of tortuosity in porous media. Despite the progress in characterization, there still exist many controversies, misconceptions, and confusions about tortuosity, which mainly come from the fact that the awareness for systematic differences between tortuosity types is missing. Hence, in many studies and discussions, the different tortuosity types are not clearly distinguished and addressed. As a first step toward solving this problem, we proposed to use a clear terminology. For this purpose, a new classification scheme and a new tortuosity-nomenclature were introduced in Chap. 2 (see e.g., Fig. 2.8). As a second step, the nature, and the extent of the systematic differences between the various tortuosity types need to be documented and illustrated. In order to investigate these differences, a large *collection of empirical data from 69 different references* was presented and analysed in this chapter (see Table 3.1).

Based on the collection of empirical data from literature, *plots of tortuosity (τ) versus porosity (ε)* are presented *for different classes of materials and microstructures*. For simple microstructures such as monosized sphere packings, the $\tau - \varepsilon$ plot shows a characteristic field that defines a relatively narrow band close to the Bruggeman trend line. However, for most material classes, which have more complicated, disordered, and stochastic microstructures (such as fuel cell electrodes, foams, rocks, and soils), the corresponding characteristic fields in the $\tau - \varepsilon$ plots are expanded over much wider domains. In addition, the characteristic fields for these materials classes are strongly overlapping. This overlap is observed even for material classes with significantly different microstructural architectures (e.g., cellular foams, fibrous textiles or sintered ceramics fabricated from powders). Obviously, *the different materials and microstructures cannot be distinguished easily based on their $\tau - \varepsilon$ characteristics.*

For the same collection of empirical data, $\tau - \varepsilon$ *characteristics are replotted*, but this time separately *for the different types of tortuosities*. Thereby, *systematic differences can be observed among the main tortuosity categories*: The indirect tortuosities show relatively high tortuosity values (typically higher than 2 and sometimes up to 20) and their $\tau - \varepsilon$ characteristics scatter over a wide range. In contrast, $\tau - \varepsilon$ characteristics of the direct geometric tortuosities and of the mixed tortuosities are typically concentrated in a narrow band close to the Bruggeman trend line (i.e., $\tau <$ 1.5 for $\varepsilon > 0.3$).

Systematic differences are also observed between tortuosity types *within the same tortuosity category*. For example, within the mixed category, the hydraulic tortuosity is consistently higher than the electrical and diffusional tortuosities. Similarly, within the category of direct geometric tortuosities, the medial axis tortuosity is consistently higher than the geodesic type. The $\tau - \varepsilon$ plots of empirical data thus document a *relative order among the different tortuosity types*. This consistent pattern is graphically illustrated in Fig. 3.9.

Hence, a *consistent $\tau - \varepsilon$ pattern* is observed when the empirical data is plotted *for different tortuosity types*. However, when the same data is plotted for different materials and microstructures, the corresponding pattern of the characteristic fields is less clear. *It follows that the tortuosity value that is measured for a specific material, is much more dependent on the type of tortuosity than it is dependent on the material and its microstructure.* This illustrates the need for a clear distinction between the different tortuosity types, including the need for a careful selection of a suitable method and the use of a clear terminology (i.e., nomenclature).

Based on the detailed description of the underlying definitions (see Chap. 2) and based on the documentation of the characteristic $\tau - \varepsilon$ patterns (present chapter), the *main characteristics of the three main tortuosity categories* can be summarized as follows:

The direct tortuosities (geodesic, FMM, medial axis, PTM, etc.) as well as the *mixed tortuosities* (streamline and volume-averaged) are *based on geometric analyses* and therefore, *they provide realistic estimations of the true path lengths*. The differences among these tortuosity types are relatively small, and they reflect the existing variations of the underlying geometric and methodological concepts.

The indirect tortuosities are *derived from effective (or relative) properties* that are known from experiment or simulation. The indirect tortuosities can thus be considered as a measure for the *bulk microstructure resistance against transport*, which includes also other resistive effects such as the bottleneck effect. Indirect tortuosities should therefore not be misinterpreted as a realistic measure for the length of transport pathways, but they should be rather considered as a *bulk factor or fudge factor*, which describes the ratio of relative transport property over porosity (e.g., $\tau_{indir\_ele} = \sqrt{(\sigma_{rel}/\varepsilon)}$).

Finally, the empirical data also illustrates that tortuosity is not strictly bound to porosity. As the pore volume decreases, the more scattering of tortuosity values can be observed. Consequently, any mathematical expression that aims to provide a generalized description of $\tau - \varepsilon$ relationships in porous media must be treated with caution (especially in cases without specification of the corresponding type of tortuosity).

References

1. K. Grew, A.A. Peracchio, J.R. Izzo, W. Chiu, Nondestructive imaging and analysis of transport processes in the solid oxide fuel cell anode, in *ECS Transactions*, vol. 25 (ECS, 2009), pp. 1861–1870
2. K.N. Grew, A.A. Peracchio, A.S. Joshi, J.R. Izzo Jr., W.K.S. Chiu, Characterization and analysis methods for the examination of the heterogeneous solid oxide fuel cell electrode microstructure. Part 1: volumetric measurements of the heterogeneous structure. J. Power Sources **195**, 7930 (2010)
3. H. Iwai et al., Quantification of SOFC anode microstructure based on dual beam FIB-SEM technique. J. Power Sources **195**, 955 (2010)
4. M. Kishimoto, H. Iwai, M. Saito, H. Yoshida, Quantitative evaluation of solid oxide fuel cell porous anode microstructure based on focused ion beam and scanning electron microscope technique and prediction of anode overpotentials. J. Power Sources **196**, 4555 (2011)
5. S.J. Cooper, M. Kishimoto, F. Tariq, R.S. Bradley, A.J. Marquis, N.P. Brandon, J.A. Kilner, P.R. Shearing, Microstructural analysis of an LSCF cathode using in situ tomography and simulation. ECS Trans. **57**, 2671 (2013)
6. S.J. Cooper, A. Bertei, P.R. Shearing, J.A. Kilner, N.P. Brandon, TauFactor: an open-source application for calculating tortuosity factors from tomographic data. SoftwareX **5**, 203 (2016)
7. J.R. Wilson, W. Kobsiriphat, R. Mendoza, H.-Y. Chen, J.M. Hiller, D.J. Miller, K. Thornton, P.W. Voorhees, S.B. Adler, S.A Barnett, Three-dimensional reconstruction of a solid-oxide fuel-cell anode. Nat. Mater. **5**, 541 (2006)

8. B. Tjaden, J. Lane, P.J. Withers, R.S. Bradley, D.J.L. Brett, P.R. Shearing, The application of 3D imaging techniques, simulation and diffusion experiments to explore transport properties in porous oxygen transport membrane support materials. Solid State Ion **288**, 315 (2016)

9. J. Joos, M. Ender, T. Carraro, A. Weber, E. Ivers-Tiffée, Representative volume element size for accurate solid oxide fuel cell cathode reconstructions from focused ion beam tomography data. Electrochim. Acta **82**, 268 (2012)

10. J. Laurencin, R. Quey, G. Delette, H. Suhonen, P. Cloetens, P. Bleuet, Characterisation of solid oxide fuel cell Ni–8YSZ substrate by synchrotron X-ray nano-tomography: from 3D reconstruction to microstructure quantification. J. Power Sources **198**, 182 (2012)

11. L. Holzer, B. Iwanschitz, Th. Hocker, L. Keller, O. Pecho, G. Sartoris, Ph. Gasser, B. Muench, Redox cycling of Ni–YSZ anodes for solid oxide fuel cells: influence of tortuosity, constriction and percolation factors on the effective transport properties. J. Power Sources **242**, 179 (2013)

12. O. Pecho, O. Stenzel, B. Iwanschitz, P. Gasser, M. Neumann, V. Schmidt, M. Prestat, T. Hocker, R. Flatt, L. Holzer, 3D microstructure effects in Ni-YSZ anodes: prediction of effective transport properties and optimization of Redox stability. Materials **8**, 5554 (2015)

13. L. Holzer, D. Wiedenmann, B. Münch, L. Keller, M. Prestat, P. Gasser, I. Robertson, B. Grobéty, The influence of constrictivity on the effective transport properties of porous layers in electrolysis and fuel cells. J. Mater. Sci. **48**, 2934 (2013)

14. D. Wiedenmann et al., Three-dimensional pore structure and ion conductivity of porous ceramic diaphragms. AIChE J. **59**, 1446 (2013)

15. K. Zheng, Y. Zhang, L. Li, M. Ni, On the tortuosity factor of solid phase in solid oxide fuel cell electrodes. Int. J. Hydrogen Energy **40**, 665 (2015)

16. A.Z. Lichtner, D. Jauffrès, D. Roussel, F. Charlot, C.L. Martin, R.K. Bordia, Dispersion, connectivity and tortuosity of hierarchical porosity composite SOFC cathodes prepared by freeze-casting. J. Eur. Ceram. Soc. **35**, 585 (2015)

17. L. Almar, J. Szasz, A. Weber, E. Ivers-Tiffée, Oxygen transport kinetics of mixed ionic-electronic conductors by coupling focused ion beam tomography and electrochemical impedance spectroscopy. J. Electrochem. Soc. **164**, F289 (2017)

18. C. Endler-Schuck, J. Joos, C. Niedrig, A. Weber, E. Ivers-Tiffée, The chemical oxygen surface exchange and bulk diffusion coefficient determined by impedance spectroscopy of porous $La_{0.58}Sr_{0.4}Co_{0.2}Fe_{0.8}O_3 - \delta$ (LSCF) cathodes. Solid State Ion **269**, 67 (2015)

19. L. Holzer et al., Fundamental relationships between 3D pore topology, electrolyte conduction and flow properties: towards knowledge-based design of ceramic diaphragms for sensor applications. Mater. Des. **99**, 314 (2016)

20. S. Haj Ibrahim, M. Neumann, F. Klingner, V. Schmidt, T. Wejrzanowski, Analysis of the 3D microstructure of tape-cast open-porous materials via a combination of experiments and modeling. Mater. Des. **133**, 216 (2017)

21. M. Kishimoto, M. Lomberg, E. Ruiz-Trejo, N.P. Brandon, Numerical modeling of nickel-infiltrated gadolinium-doped ceria electrodes reconstructed with focused ion beam tomography. Electrochim. Acta **190**, 178 (2016)

22. N.O. Shanti, V.W.L. Chan, S.R. Stock, F. De Carlo, K. Thornton, K.T. Faber, X-ray micro-computed tomography and tortuosity calculations of percolating pore networks. Acta Mater. **71**, 126 (2014)

23. Z. Yu, R.N. Carter, J. Zhang, Measurements of pore size distribution, porosity, effective oxygen diffusivity, and tortuosity of PEM fuel cell electrodes. Fuel Cells **12**, 557 (2012)

24. J. Sarkar, S. Bhattacharyya, Application of graphene and graphene-based materials in clean energy-related devices Minghui. Arch. Thermodyn. **33**, 23 (2012)

25. P.A. García-Salaberri, J.T. Gostick, G. Hwang, A.Z. Weber, M. Vera, Effective diffusivity in partially-saturated carbon-fiber gas diffusion layers: effect of local saturation and application to macroscopic continuum models. J. Power Sources **296**, 440 (2015)

26. D. Froning, J. Yu, G. Gaiselmann, U. Reimer, I. Manke, V. Schmidt, W. Lehnert, Impact of compression on gas transport in non-woven gas diffusion layers of high temperature polymer electrolyte fuel cells. J. Power Sources **318**, 26 (2016)

27. R. Flückiger, S.A. Freunberger, D. Kramer, A. Wokaun, G.G. Scherer, F.N. Büchi, Anisotropic, effective diffusivity of porous gas diffusion layer materials for PEFC. Electrochim. Acta **54**, 551 (2008)

28. D. Froning, J. Brinkmann, U. Reimer, V. Schmidt, W. Lehnert, D. Stolten, 3D analysis, modeling and simulation of transport processes in compressed fibrous microstructures, using the lattice Boltzmann method. Electrochim. Acta **110**, 325 (2013)

29. L. Holzer, O. Pecho, J. Schumacher, Ph. Marmet, O. Stenzel, F.N. Büchi, A. Lamibrac, B. Münch, Microstructure-Property Relationships in a Gas Diffusion Layer (GDL) for Polymer Electrolyte Fuel Cells, Part I: Effect of Compression and Anisotropy of Dry GDL. Electrochim. Acta **227**, 419 (2017)

30. L. Holzer, O. Pecho, J. Schumacher, Ph. Marmet, F. N. Büchi, A. Lamibrac, B. Münch, Microstructure-property relationships in a gas diffusion layer (GDL) for polymer electrolyte fuel cells, part II: pressure-induced water injection and liquid permeability. Electrochim. Acta **241**, 414 (2017)

31. X. Huang, Q. Wang, W. Zhou, D. Deng, Y. Zhao, D. Wen, J. Li, Morphology and transport properties of fibrous porous media. Powder Technol. **283**, 618 (2015)

32. S.J. Cooper et al., Image based modelling of microstructural heterogeneity in $LiFePO_4$ electrodes for Li-ion batteries. J. Power Sources **247**, 1033 (2014)

33. F. Tariq, V. Yufit, M. Kishimoto, P.R. Shearing, S. Menkin, D. Golodnitsky, J. Gelb, E. Peled, N.P. Brandon, Three-dimensional high resolution X-ray imaging and quantification of lithium ion battery mesocarbon microbead anodes. J. Power Sources **248**, 1014 (2014)

34. P.R. Shearing, N.P. Brandon, J. Gelb, R. Bradley, P.J. Withers, A.J. Marquis, S. Cooper, S.J. Harris, Multi length scale microstructural investigations of a commercially available Li-ion battery electrode. J. Electrochem. Soc. **159**, A1023 (2012)

35. O.O. Taiwo, D.P. Finegan, D.S. Eastwood, J.L. Fife, L.D. Brown, J.A. Darr, P.D. Lee, D.J.L. Brett, P.R. Shearing, Comparison of three-dimensional analysis and stereological techniques for quantifying lithium-ion battery electrode microstructures. J. Microsc. **263**, 280 (2016)

36. L. Almar, J. Joos, A. Weber, E. Ivers-Tiffée, Microstructural feature analysis of commercial Li-ion battery cathodes by focused ion beam tomography. J. Power Sources **427**, 1 (2019)

37. P.R. Shearing, L.E. Howard, P.S. Jørgensen, N.P. Brandon, S.J. Harris, Characterization of the 3-dimensional microstructure of a graphite negative electrode from a Li-ion battery. Electrochem. Commun. **12**, 374 (2010)

38. T. Hutzenlaub, A. Asthana, J. Becker, D.R. Wheeler, R. Zengerle, S. Thiele, FIB/SEM-based calculation of tortuosity in a porous LiCoO2 cathode for a Li-ion battery. Electrochem. Commun. **27**, 77 (2013)

39. M. Ebner, D.W. Chung, R.E. García, V. Wood, Tortuosity anisotropy in lithium-ion battery electrodes. Adv. Energy Mater. **4**, 1 (2014)

40. T. Hamann, L. Zhang, Y. Gong, G. Godbey, J. Gritton, D. McOwen, G. Hitz, E. Wachsman, The effects of constriction factor and geometric tortuosity on Li-ion transport in porous solid-state Li-ion electrolytes. Adv. Funct. Mater. **30** (2020)

41. J. Landesfeind, A. Eldiven, H.A. Gasteiger, Influence of the binder on lithium ion battery electrode tortuosity and performance. J. Electrochem. Soc. **165**, A1122 (2018)

42. J. Landesfeind, M. Ebner, A. Eldiven, V. Wood, H.A. Gasteiger, Tortuosity of battery electrodes: validation of impedance-derived values and critical comparison with 3D tomography. J. Electrochem. Soc. **165**, A469 (2018)

43. J. Landesfeind, J. Hattendorff, A. Ehrl, W.A. Wall, H.A. Gasteiger, Tortuosity determination of battery electrodes and separators by impedance spectroscopy. J. Electrochem. Soc. **163**, A1373 (2016)

44. C.F. Berg, Permeability description by characteristic length, tortuosity, constriction and porosity. Transp. Porous Media **103**, 381 (2014)

45. J.L. Provis, R.J. Myers, C.E. White, V. Rose, J.S.J. Van Deventer, X-ray microtomography shows pore structure and tortuosity in alkali-activated binders. Cem. Concr. Res. **42**, 855 (2012)

46. L.J. Klinkenberg, Analogy between diffusion and elecrical conductivity in porous rocks. GSA Bull. **62**, 559 (1951)

47. A.S. Ziarani, R. Aguilera, Pore-throat radius and tortuosity estimation from formation resistivity data for tight-gas sandstone reservoirs. J. Appl Geophy **83**, 65 (2012)

48. L.M. Keller, L. Holzer, R. Wepf, P. Gasser, 3D geometry and topology of pore pathways in Opalinus clay: implications for mass transport. Appl Clay Sci **52**, 85 (2011)

49. C.F. Berg, Re-examining Archie's law: conductance description by tortuosity and constriction. Phys. Rev. E **86**, 046314 (2012)

50. L.M. Keller, L. Holzer, R. Wepf, P. Gasser, B. Münch, P. Marschall, On the application of focused ion beam nanotomography in characterizing the 3D pore space geometry of Opalinus clay. Phys. Chem. Earth **36**, 1539 (2011)

51. A.H. Kristensen, A. Thorbjørn, M.P. Jensen, M. Pedersen, P. Moldrup, Gas-phase diffusivity and tortuosity of structured soils. J. Contam. Hydrol. **115**, 26 (2010)

52. T.B. Boving, P. Grathwohl, Tracer diffusion coefficients in sedimentary rocks: correlation to porosity and hydraulic conductivity. J. Contam. Hydrol. **53**, 85 (2001)

53. R. Nemati, J. Rahbar Shahrouzi, R. Alizadeh, A stochastic approach for predicting tortuosity in porous media via pore network modeling. Comput. Geotech. **120**, 103406 (2020)

54. H. Saomoto, J. Katagiri, Direct comparison of hydraulic tortuosity and electric tortuosity based on finite element analysis. Theor. Appl. Mech. Lett. **5**, 177 (2015)

55. H. Saomoto, J. Katagiri, Particle shape effects on hydraulic and electric tortuosities: a novel empirical tortuosity model based on van Genuchten-type function. Transp. Porous Media **107**, 781 (2015)

56. A. Nabovati, A.C.M. Sousa, Fluid flow simulation in random porous media at pore level using the lattice Boltzmann method. Sci. Technol. **2**, 226 (2007)

57. A. Ghassemi, A. Pak, Pore scale study of permeability and tortuosity for flow through particulate media using lattice Boltzmann method. Int. J. Numer. Anal. Methods Geomech. **35**, 886 (2011)

58. C.J. Gommes, A.-J. Bons, S. Blacher, J.H. Dunsmuir, A.H. Tsou, Practical methods for measuring the tortuosity of porous materials from binary or gray-tone tomographic reconstructions. AIChE J. **55**, 2000 (2009)

59. D.W. Chung, M. Ebner, D.R. Ely, V. Wood, R. Edwin García, Validity of the Bruggeman relation for porous electrodes. Model. Simul. Mat. Sci. Eng. **21**, (2013)

60. W. Sobieski, The use of path tracking method for determining the tortuosity field in a porous bed. Granul. Matter **18**, 1 (2016)

61. B. Sheikh, A. Pak, Numerical investigation of the effects of porosity and tortuosity on soil permeability using coupled three-dimensional discrete-element method and lattice Boltzmann method. Phys. Rev. E Stat. Nonlin. Soft Matter Phys. **91**, 1 (2015)

62. S. Pawlowski, N. Nayak, M. Meireles, C.A.M. Portugal, S. Velizarov, J.G. Crespo, CFD modelling of flow patterns, tortuosity and residence time distribution in monolithic porous columns reconstructed from X-ray tomography data. Chem. Eng. J. **350**, 757 (2018)

63. R.I. Al-Raoush, I.T. Madhoun, TORT3D: a MATLAB code to compute geometric tortuosity from 3D images of unconsolidated porous media. Powder Technol. **320**, 99 (2017)

64. K. Hormann, V. Baranau, D. Hlushkou, A. Höltzel, U. Tallarek, Topological analysis of non-granular, disordered porous media: determination of pore connectivity, pore coordination, and geometric tortuosity in physically reconstructed silica monoliths. New J. Chem. **40**, 4187 (2016)

65. T.F. Johnson, F. Iacoviello, D.J. Hayden, J.H. Welsh, P.R. Levison, P.R. Shearing, D.G. Bracewell, Packed bed compression visualisation and flow simulation using an erosion-dilation approach. J. Chromatogr. A **1611**, 1 (2020)

66. O. Stenzel, O. Pecho, L. Holzer, M. Neumann, V. Schmidt, predicting effective conductivities based on geometric microstructure characteristics. AIChE J. **62**, 1834 (2016)

67. G. Gaiselmann, M. Neumann, V. Schmidt, O. Pecho, T. Hocker, L. Holzer, Quantitative relationships between microstructure and effective transport properties based on virtual materials testing. AIChE J. **60**, 1983 (2014)

68. M.A. Knackstedt, C.H. Arns, M. Saadatfar, T.J. Senden, A. Sakellariou, A.P. Sheppard, R.M. Sok, W. Schrof, H. Steininger, Virtual materials design: properties of cellular solids derived from 3D tomographic images. Adv. Eng. Mater. **7**, 238 (2005)

69. J. Vicente, F. Topin, J.V. Daurelle, Open celled material structural properties measurement: from morphology to transport properties. Mater. Trans. **47**, 2195 (2006)

70. A. Rohatgi, *Webplotdigitizer*. https://automeris.io/WebPlotDigitizer/

71. M.B. Clennell, Tortuosity: a guide through the maze, in *Developments In Petrophysics*, ed by M.A. Lovell, P.K. Harvey (Geol. Soc. Spec. Publ. No. 122, 1997), pp. 299–344

72. J. Fu, H.R. Thomas, C. Li, Tortuosity of porous media: image analysis and physical simulation. Earth Sci. Rev. **1** (2020)

73. L. Shen, Z. Chen, Critical review of the impact of tortuosity on diffusion. Chem. Eng. Sci. **62**, 3748 (2007)

74. B. Ghanbarian, A.G. Hunt, R.P. Ewing, M. Sahimi, Tortuosity in porous media: a critical review. Soil Sci. Soc. Am. J. **77**, 1461 (2013)

75. B. Tjaden, D.J.L. Brett, P.R. Shearing, Tortuosity in electrochemical devices: a review of calculation approaches. Int. Mater. Rev. **63**, 47 (2018)

76. A. Idris, A. Muntean, B. Mesic, A review on predictive tortuosity models for composite films in gas barrier applications. J. Coat Technol. Res. (2022)

77. M. Neumann, O. Stenzel, F. Willot, L. Holzer, V. Schmidt, Quantifying the influence of microstructure on effective conductivity and permeability: virtual materials testing. Int. J. Solids Struct. **184**, 211 (2020)

78. O. Stenzel, O. Pecho, L. Holzer, M. Neumann, V. Schmidt, Big Data for microstructure-property relationships: a case study of predicting effective conductivities. AIChE J. **63**, 4224 (2017)

79. G.E. Archie, The electrical resistivity log as an aid in determining some reservoir characteristics. Trans. AIME **146** (1942)

80. M.R.J. Wyllie, W.D. Rose, Some theoretical considerations related to the quantitative evaluation of the physical characteristics of reservoir rock from electrical log data. J. Petrol. Technol. **2**, 105 (1950)

81. B. Yu, J. Li, Some fractal characters of porous media. Fractals **9**, 365 (2001)

82. P.-Y. Lanfrey, Z.V. Kuzeljevic, M.P. Dudukovic, Tortuosity model for fixed beds randomly packed with identical particles. Chem. Eng. Sci. **65**, 1891 (2010)

83. J.-H. Li, B.-M. Yu, Tortuosity of flow paths through a Sierpinski Carpet. Chin. Phys. Lett. **28**, 034701 (2011)

84. J. Comiti, M. Renaud, A new model for determining mean structure parameters of fixed beds from pressure drop measurements: application to beds packed with parallelepipedal particles. Chem. Eng. Sci. **44**, 1539 (1989)

85. M. Mota, J.A. Teixeira, A. Yelshin, Binary spherical particle mixed beds porosity and permeability relationship measurement. Trans. Filtr. Soc. **1**, 101 (2001)

86. J.P. Du Plessis, J.H. Masliyah, Flow through isotropic granular porous media. Transp. Porous Media **6**, 207 (1991)

87. M.M. Ahmadi, S. Mohammadi, A.N. Hayati, Analytical derivation of tortuosity and permeability of monosized spheres: a volume averaging approach. Phys. Rev. E **83**, 026312 (2011)

88. A. Koponen, M. Kataja, J. Timonen, Tortuous flow in porous media. Phys. Rev. E **54**, 406 (1996)

89. J.C. Maxwell, *A Treatise on Electricity and Magnetism*, vol. I (Clarendon Press, London, 1873)

90. J.M.P.Q. Delgado, A simple experimental technique to measure tortuosity in packed beds. Can. J. Chem. Eng. **84**, 651 (2008)

91. J.W. Beeckman, Mathematical description of heterogeneous materials. Chem. Eng. Sci. **45**, 2603 (1990)

92. N. Iversen, B.B. Jørgensen, Diffusion coefficients of sulfate and methane in marine sediments: influence of porosity. Geochim. Cosmochim. Acta **57**, 571 (1993)

93. E.E. Petersen, Diffusion in a pore of varying cross section. AIChE J. **4**, 343 (1958)
94. D.A.G. Bruggeman, Berechnung Verschiedener Physikalischer Konstanten von Heterogenen Substanzen. I. Dielektrizitätskonstanten Und Leitfähigkeiten Der Mischkörper Aus Isotropen Substanzen. Ann. Phys. **416**, 636 (1935)
95. B. Tjaden, S.J. Cooper, D.J. Brett, D. Kramer, P.R. Shearing, On the origin and application of the Bruggeman correlation for analysing transport phenomena in electrochemical systems. Curr. Opin. Chem. Eng. **12**, 44 (2016)
96. I.V. Thorat, D.E. Stephenson, N.A. Zacharias, K. Zaghib, J.N. Harb, D.R. Wheeler, Quantifying tortuosity in porous Li-ion battery materials. J. Power Sources **188**, 592 (2009)
97. J. Van Brakel, P.M. Heertjes, Analysis of diffusion in macroporous media in terms of a porosity, a tortuosity and a constrictivity factor. Int. J. Heat Mass Transf. 1093 (1974)
98. M.M. Tomadakis, S.V. Sotirchos, Transport through random arrays of conductive cylinders dispersed in a conductive matrix. J. Chem. Phys. **104**, 6893 (1996)
99. H.L. Weissberg, Effective diffusion coefficient in porous media. J. Appl. Phys. **34**, 2636 (1963)
100. Z. Sun, X. Tang, G. Cheng, Numerical simulation for tortuosity of porous media. Microporous Mesoporous Mater. **173**, 37 (2013)
101. J.S. Mackie, P. Meares, The diffusion of electrolytes in a cation-exchange resin membrane. Proc. R. Soc. A **232**, 498 (1955)
102. L. Pisani, Simple expression for the tortuosity of porous media. Transp. Porous Media **88**, 193 (2011)
103. Y. Jin, J.B. Dong, X. Li, Y. Wu, Kinematical measurement of hydraulic tortuosity of fluid flow in porous media. Int. J. Mod. Phys. C **26** (2015)
104. A. Duda, Z. Koza, M. Matyka, Hydraulic tortuosity in arbitrary porous media flow. Phys. Rev. E Stat. Nonlin. Soft Matter Phys. **84**, (2011)
105. M. Ebner, V. Wood, Tool for tortuosity estimation in lithium ion battery porous electrodes. J. Electrochem. Soc. **162**, A3064 (2015)
106. M. Doyle, Comparison of modeling predictions with experimental data from plastic lithium ion cells. J. Electrochem. Soc. **143**, 1890 (1996)
107. P. Arora, M. Doyle, A.S. Gozdz, R.E. White, J. Newman, Comparison between computer simulations and experimental data for high-rate discharges of plastic lithium-ion batteries. J. Power Sources **88**, 219 (2000)
108. D. Kehrwald, P.R. Shearing, N.P. Brandon, P.K. Sinha, S.J. Harris, Local tortuosity inhomogeneities in a lithium battery composite electrode. J. Electrochem. Soc. **158**, A1393 (2011)
109. J. Cannarella, C.B. Arnold, Ion transport restriction in mechanically strained separator membranes. J. Power Sources **226**, 149 (2013)
110. N.A. Zacharias, D.R. Nevers, C. Skelton, K. Knackstedt, D.E. Stephenson, D.R. Wheeler, Direct measurements of effective ionic transport in porous Li-ion electrodes. J. Electrochem. Soc. **160**, A306 (2013)

Chapter 4
Image Based Methodologies, Workflows, and Calculation Approaches for Tortuosity

Abstract In this chapter, modern methodologies for characterization of tortuosity are thoroughly reviewed. Thereby, 3D microstructure data is considered as the most relevant basis for characterization of all three tortuosity categories, i.e., direct geometric, indirect physics-based and mixed tortuosities. The workflows for tortuosity characterization consists of the following methodological steps, which are discussed in great detail: (a) 3D imaging (X-ray tomography, FIB-SEM tomography and serial sectioning, Electron tomography and atom probe tomography), (b) qualitative image processing (3D reconstruction, filtering, segmentation) and (c) quantitative image processing (e.g., morphological analysis for determination of direct geometric tortuosity). (d) Numerical simulations are used for the estimation of effective transport properties and associated indirect physics-based tortuosities. Mixed tortuosities are determined by geometrical analysis of flow fields from numerical transport simulation. (e) Microstructure simulation by means of stochastic geometry or discrete element modeling enables the efficient creation of numerous virtual 3D microstructure models, which can be used for parametric studies of micro–macro relationships (e.g., in context with digital materials design or with digital rock physics). For each of these methodologies, the underlying principles as well as the current trends in technical evolution and associated applications are reviewed. In addition, a list with 75 software packages is presented, and the corresponding options for image processing, numerical simulation and stochastic modeling are discussed. Overall, the information provided in this chapter shall help the reader to find suitable methodologies and tools that are necessary for efficient and reliable characterization of specific tortuosity types.

Supplementary Information The online version contains supplementary material available at https://doi.org/10.1007/978-3-031-30477-4_4.

4.1   Introduction

In Chap. 2, a new classification scheme with three main tortuosity categories was proposed, which includes direct geometric tortuosities, indirect physics-based tortuosities, and mixed tortuosities (see Fig. 2.8). For each of these categories, the characterization procedure has to follow a specific workflow. In our description of the methods for characterization of tortuosity we follow step by step the general workflows, which are illustrated schematically in Fig. 4.1.

For all three tortuosity categories, the modern characterization approach focuses on the collection and quantitative analysis of 3D information. Thereby, real 3D microstructure models typically originate from experimental samples that are investigated by suitable tomography methods, followed by qualitative image processing (i.e., 3D reconstruction and filtering). Alternatively, virtual 3D microstructure models can be created with methods of stochastic geometry. This approach enables the creation of a large number of virtual 3D models in an efficient way. The virtual 3D microstructure realizations are then used as a basis for parametric studies and data driven microstructure investigations.

3D microstructure models (real or virtual) are considered here as the first step in the workflows for all three tortuosity categories. For measuring direct geometric tortuosities (i.e., τ_{dir_geom}), morphological analysis of the transporting phase (typically pore phase) is then performed by means of quantitative image processing. Numerous algorithms are nowadays available (in free and in commercial software tools) for measuring of different geometric tortuosity types.

The 3D microstructure models (real or virtual) can also be used as input for numerical transport simulations, from which effective transport properties and indirect physics-based tortuosities can be derived ($\tau_{indir\_phys\_sim}$). Also here, numerous SW codes are nowadays available to simulate different transports within a 3D microstructure model (e.g., electrical or thermal conduction, bulk diffusion, Knudsen diffusion, viscous flow).

As a third option, the 3D volume fields representing the local flux from numerical transport simulations can be used as basis for the computation of mixed type tortuosities ($\tau_{mixed\_phys\_streamline}$, $\tau_{mixed\_phys\_Vav}$).

It must be noted here, that for the indirect physics-based tortuosities, there also exists an alternative way of determination without using 3D information. This alternative and complementary way is based on experimental measurement of effective transport properties ($\tau_{indir\_phys\_exp}$). However, since we consider 3D analysis as the key to a better understanding of tortuosity and associated path length effects, we are focusing in the present chapter not on the experimental approaches, but on the modern image-based methods.

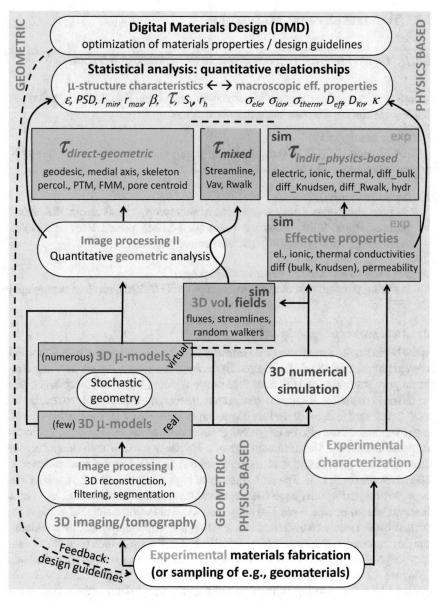

Fig. 4.1 Schematic illustration of methodologies and workflows for measuring direct geometric, indirect physics-based and mixed tortuosities. Round boxes represent methods and processes, which are discussed in Chap. 4. Rectangular boxes represent either data that is used as input for—or results that are obtained as output from the mentioned processes. Two boxes on top indicate the broader scientific context of tortuosity, which aims to establish quantitative micro–macro relationships and/ or to perform digital materials design (see Chap. 5). *Legend/abbreviations:* ε porosity, *PSD* pore size distribution, r_{min} mean bottleneck radius, r_{max} mean radius of pore bulges, β constrictivity, τ tortuosity, S_V surface area per volume, r_h hydraulic radius, σ conductivity, D diffusivity, κ permeability, *PTM* path tracking method, *FMM* fast marching method, *Vav* Volume averaged, *Rwalk* random walk, *el* electric, *diff* diffusion(-al), *hydr* hydraulic

4.2   Tomography and 3D Imaging

4.2.1   Overview and Introduction to 3D Imaging Methods

In the following sections, we consider four main categories of tomography techniques that are relevant for 3D pore-scale characterization:

(a)   ***X-ray tomography***
μ-CT, nano-CT (CT = computed tomography), transmission and scanning X-ray microscopy (TXM, SXM).

(b)   ***Serial sectioning methods***
focused ion beam - scanning electron microscopy (Dual Beam FIB-SEM), plasma (P)FIB-SEM, broad ion beam (BIB-SEM), pulsed laser, mechanical sectioning (Ultra-Microtom) and mechanical polishing.

(c)   ***3D TEM (transmission electron microscopy)***
scanning transmission electron tomography (3D STEM), electron tomography (ET).

(d)   ***Atom probe tomography (APT)***

At the beginning of a microstructure investigation there is always the question which tomography method should be chosen. To answer this question, first order criteria are the range of resolution and the size of the image window, which can be obtained with the different methods. A suitable tomography method must be capable to resolve the smallest relevant features of the investigated microstructure. At the same time, the 3D image window should also be large enough to capture the largest objects of interest in a representative way. The minimum size of a 3D image window with statistical relevance is called representative elementary volume (REV). For the determination of REV sizes see e.g., [1, 2]. The requirements of high resolution (small voxels) and at the same time, sufficiently large (i.e., representative) image window sizes are contradictory constraints, which must be addressed when choosing a suitable tomography method for materials characterization. Finding a good compromise for conflicting imaging parameters (i.e., resolution vs. REV) is a challenge, which requires a sound knowledge of the limitations and possibilities of the available tomography methods.

Figure 4.2 illustrates the range of resolutions and image window sizes that can be achieved with X-ray tomography, FIB-SEM tomography, electron tomography and atom probe tomography. Thereby the colored rectangles represent the performance fields that were typically achieved 10–15 years ago (taken from Uchic et al. [3]). At that time the different tomography methods occupied distinct performance fields (regarding resolution and image window size) with almost no overlap.

In the meanwhile, the resolution power of X-ray tomography has tremendously improved. The evolution from μCT to nanoCT is indicated with a green arrow in Fig. 4.2. For FIB-SEM tomography, the evolution went in the opposite direction. Nowadays the improvement of ion milling efficiency enables to capture much larger

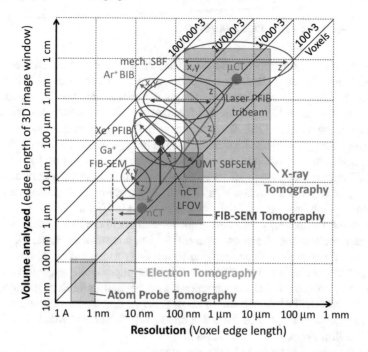

Fig. 4.2 Graphical representation of important tomography methods characterized by their typical voxel resolutions (x-axis) and size of analyzed volume (y-axis, i.e., image window size). Each diagonal line represents a specific size of data cube (i.e., constant number of voxels), if the 3D image window is isometric. The colored rectangles indicate characteristic performance fields for traditional tomography methods, which are redrawn from Uchic et al. [3]. The green arrows indicate recent methodological evolutions from μ-CT to nano-CT and to large-field-of-view (LFOV) nano-CT. Ellipsoids represent the performance of serial sectioning methods. The elliptical shapes of their performance fields result from the fact that the serial sectioning methods tend to provide anisometric data cubes, because they reveal different properties in x-, y- and z-directions. *Legend:* nCT = nano-CT, μCT = micro-CT, LFOV = large-field-of-view, PFIB = plasma FIB, UMT SBFSEM = ultra-micro-tomography serial block face SEM, BIB = broad ion beam, mech. SBF = mechanical serial block face sectioning

image windows. The evolution towards larger image windows also took place due to the introduction of new serial sectioning methods with higher milling rates (e.g., with plasma FIB and pulsed laser), whose performances are indicated with ellipses in Fig. 4.2.

In summary, the performance fields of X-ray CT, FIB-SEM tomography and other serial sectioning methods nowadays show a considerable overlap. However, it must be emphasized that the performance of a tomography method does not only depend on resolution and image window size. In particular, contrast and detection modes, acquisition time, but also the required sample properties (e.g., stability of a sample under specific imaging conditions, required sample size and required sample preparation) must be considered when choosing a suitable tomography method.

4.2.2   X-ray Computed Tomography

The resolution of X-ray tomography (XCT) has tremendously improved over the last 10 years from μm-range down to the 10 nm-range. XCT is now capable to resolve the microstructure at pore scale of almost any material in engineering science (e.g., energy materials used for batteries and fuel cells [4–20], concrete and asphalt [21, 22], polymer composites [23], 3D-printed materials [24]) as well as materials from geo-applications [25–32] and life sciences [33]. However, it must be emphasized that the progress is not restricted to the resolution power alone. Note that 10 to 15 years ago, XCT mainly was a static 3D methodology with micrometer resolution (μ-CT), which typically provided attenuation contrast. Meanwhile, the high-end version of X-ray tomography provides spatial resolutions down to 10 nm (nano-CT). It can be used as a 4D methodology with fast acquisition times in the sub-second range and it provides multi-mode detection capabilities (i.e., attenuation, phase, diffraction, and chemical contrasts). When speaking about X-ray tomography, we have thus to consider a very versatile group of 3D and 4D imaging methodologies, which continue to make fast progress in various directions, as discussed in a recent paper by Yan et al. [34]. For detailed information we also refer to the excellent review and overview articles by Cocco et al. [35], Maire and Withers [36], Pietsch and Wood [37], Brisard et al. [21], Rawson et al. [33] and Zeiss [38].

The following aspects of XCT are important:

4.2.2.1   Basic Principles of XCT

In order to understand the trends in X-ray tomography one has to consider the underlying principles at first. Figure 4.3 represents a schematic illustration of a modern CT-system. The sample is placed between the X-ray source and the detector. The X-rays penetrate the rotating sample so that a multitude of 2D projections are detected under different angles. Algorithms for 3D-reconstruction (e.g., so-called filtered back projection algorithms) are then used to reveal the internal 3D microstructure from the numerous 2D-projections. In absorption/attenuation tomography, the beam—sample interaction and associated beam intensity (I) after travelling through the sample is described by the Beer-Lambert equation

$$I = I_0 e^{-\int \mu(s)} ds, \tag{4.1}$$

where I_0 is the initial beam intensity, μ is the local attenuation coefficient and s is the beam path vector. Thereby, the attenuation coefficient (μ) is a function of electron density and atomic number (Z). Local variations of this attenuation coefficient are directly related to the materials microstructure, which can be reconstructed in 3D (e.g., by means of filtered back projection). Performance limitations such as the acquisition rates, maximum sample size or inherent noise, can be understood from these basic methodological principles.

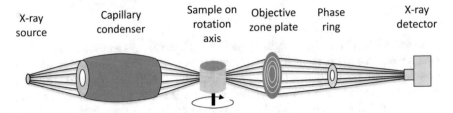

Fig. 4.3 Schematic illustration of a modern, laboratory-based X-ray tomography system, redrawn from Zeiss [38]

Fig. 4.4 Schematic illustration of imaging parameters, which need to be considered, when optimizing image acquisition with X-ray computed tomography (XCT) and/or X-ray microscopy (XM)

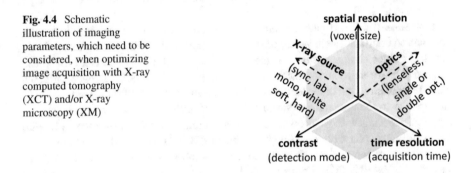

As indicated in Fig. 4.4, spatial resolution is only one of the performance relevant characteristics. Other important performance characteristics are for example time resolution/acquisition time and contrast/detection modes. The type of the X-ray source and the optical system also has a strong impact on the performance. Tremendous progress was achieved in all these technological fields. However, there also exists a multitude of interdependencies between the characteristics and parameters mentioned above. For example, faster acquisition usually leads to higher noise and therefore also to weaker contrast and lower effective spatial resolution. In the following, we briefly discuss the most important performance parameters and associated interdependencies.

4.2.2.2 Attenuation and Phase Contrasts

The interaction of X-rays with the material depends on the complex refractive index (n) with a real part ($1-\delta$) and an imaginary part (β_{x-ray}), i.e.,

$$n = (1 - \delta) + i\beta_{x-ray}. \tag{4.2}$$

The attenuation coefficient (μ) itself depends on the imaginary part (β_{x-ray}) of the complex refractive index and on the wavelength (λ), where

$$\mu = 4\pi \beta_{x-ray}/\lambda. \qquad (4.3)$$

In materials with a high imaginary part ($\beta_{x\text{-}ray}$), the X-rays amplitude is damped, which leads to a lower intensity (i.e., stronger absorption). The attenuation strongly decreases with the beam energy E. More precisely, it holds that $\beta_{x\text{-}ray} = 1/E^4$. Soft X-rays thus provide a better attenuation contrast and also a higher resolution can be achieved. However, stronger absorption at lower beam energy limits the size of samples that can be transmitted and at the same time increases the problem of beam damage with soft X-rays. The contrast between different material constituents can also be improved significantly by using dual-energy X-ray tomography (see e.g., Gondzio et al. [39]).

Phase contrast (PC) imaging is an interesting alternative for materials with a weak attenuation contrast. Local variations of the real part of the refractive index ($1-\delta$, see Eq. 4.2) induce changes of the wavelengths, which lead to beam deflections. These refractive beam-material interactions can be indirectly detected as a phase shift (ϕ). The 3D reconstruction of the local δ-value reveals the materials phase contrast, which, in opposite to the attenuation contrast, increases with beam energy ($\Delta\delta/\Delta\beta_{x\text{-}ray} = E^2$). For materials with weak absorption contrast (e.g., in battery electrodes: graphite versus lithium), phase contrast imaging with hard X-rays usually gives better results.

There are different methods for phase contrast imaging (propagation PC, grating-based PC, Zernike PC). These contrast modes require more complex optics, a coherent beam source and more sophisticated 3D reconstructions. However, nowadays even lab-based tomography systems offer the option of phase contrast imaging. For further explanations on attenuation and phase contrast as well as on the principles of 3D reconstruction we refer to Pietsch and Wood [37] and the references therein.

4.2.2.3   Spatial Resolution and Magnification

In principle, X-ray microscopy offers three types of magnifications, which are based on X-ray optics, light optics or geometric set-up.

(a) **X-ray optics**

X-ray optics work with reflection, diffraction, or refraction (see e.g., www.x-ray-optics.de). Recent progress in nanofabrication has boosted the technology for X-ray optics (e.g., Fresnel Zone Plates FZP), which now enables *voxel resolutions down to 10 nm*. A *major drawback of X-ray optics* is the fact that it generally reduces the beam flux and therefore leads to *longer acquisition times*. This can be partially compensated by using brilliant synchrotron sources with a high beam flux.

(b) **Light optics**

Conventional light optics can be introduced in detection systems that consist of a scintillator and a CCD or CMOS camera. The scintillator converts the X-rays into visible light, which can then be magnified with optical lenses similar to conventional light microscopes. In contrast to X-ray optics, the light optics is highly efficient and fast. However, the diffraction limit of visible light constrains the resolution to little less than 1 μm.

(c) **Geometric set-up**

In systems with divergent, cone-shaped beams, the image can be geometrically magnified by adjusting the relative positions of source, object, and detector (see Fig. 4.3). The resolution in lens-less systems with purely geometric magnification is typically limited to the μm-range.

In modern X-ray tomography systems these three magnifying methods are combined, so that *a high magnification* can be achieved together with a relatively high efficiency and a relatively *fast acquisition time* (see e.g., [38]). The evolution of 'best' spatial resolution in X-ray tomography over the last 50 years is illustrated in Fig. 4.5 a (adapted from Maire and Withers [36]). This figure shows that it is now possible to reach voxel resolutions in the 10 nm-range by soft X-ray tomography at synchrotron beam lines equipped with Fresnel Zone Plate optics. The resolution power for hard X-rays is weaker. However, soft X-ray tomography suffers from the limited sample thickness, especially for materials with strong absorption (e.g., metals, heavy elements, high density).

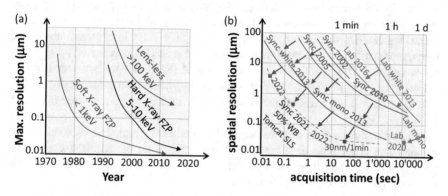

Fig. 4.5 Improvement of the spatial resolution in X-ray tomography. **a** Evolution of best resolution over the last 50 years for lens-less systems (blue), for hard X-ray tomography with Fresnel Zone Plate (FZP) (black) and for soft X-ray tomography with FZP (red). **b** Evolution of X-ray tomography, illustrating the link between spatial and temporal resolutions. Red: Synchrotron, white beam / Blue: Synchrotron, monochromatic / Green: Laboratory systems. The violet point marks the current high-end performance with sub-second and sub-μm resolutions achieved at synchrotron with 50% white + 50% monochromatic light (pers. communication, F. Büchi, PSI, Swiss Light Source SLS, 2021). The curves in **a** and **b** are redrawn from Maire and Withers [36] and updated with current trends presented in the literature. For recent progress in nano-CT see Yan et al. [34]

4.2.2.4   Time Resolution Versus Spatial Resolution

In general, a fast acquisition time comes at the expense of increased noise. Synchrotron beams with a high brilliance can reduce the noise and are thus particularly well suited for fast X-ray imaging. Nowadays, 4D imaging at 20 to 50 Hz has become possible with white beam synchrotron tomography (see e.g., Maire et al. [40]). As shown in Fig. 4.5b, time resolution and spatial resolution are constraining each other. Both of them also strongly depend on various other aspects such as the facility type (synchrotron versus laboratory systems), beam energy and beam intensity (monochromatic versus white versus mixed beams). Also, the attenuation contrast of the material under investigation and the size of the sample are representing constraining factors.

The best time resolution can be achieved with a white beam (WB) of synchrotron sources due to the relatively high beam flux, however with certain limitations in magnification. Monochromatic beams are better suited to exploit the power of magnifying systems such as FZP, which opens new capabilities for nano-CT [34]. Even lab-based systems with X-ray optics are nowadays capable of providing 50 nm resolutions. However, lab-based systems exhibit a relatively long acquisition time, which is typically several hours per 3D image. A good compromise between short acquisition time, high spatial resolution and reasonable signal to noise ratio can be reached by mixing monochromatic and white beams (e.g., 50% WB). In this way, sub-second tomography with sub-μm resolution has become possible, as reported by F. Büchi (pers. comm., 2021) for the Swiss Light Source (SLS) at Paul Scherrer Institute (violet data point in Fig. 4.5b). For white beam synchrotron tomography, 30 nm spatial resolutions at 1 min acquisition time are reported as current state of the art (Yan et al. [34]).

Many parameters must be considered when optimizing temporal vs. spatial resolutions. The acquisition time can be estimated based on the required number of projections (N), which increases with the image window size and associated number of pixels (q) in horizontal direction, i.e.,

$$N = q\pi/2. \tag{4.4}$$

For example, for $q = 1024$ pixels, the required number of 2D projections (N) is equal to 1570. With a total rotation of 180° this results in 8.7 projections per 1°. For these settings, 3D imaging at 1 Hz (i.e., 1570 projections per second) requires an acquisition time of 0.64 ms for a single 2D projection. By pushing the limits of fast tomography towards 100 Hz (e.g., 157,000 projections per second), new technological solutions needed to be developed such as fast signal processing (e.g., read out and storage of up to 100 GB per second, see Mokso et al. [41]), highly efficient optics and detector systems (Bührer et al. [19]), more brilliant sources, intelligent 3D reconstruction strategies (e.g., based on a smaller number of projections or segmentation oriented reconstructions using a priori knowledge about the phases, see [42–45]). Finally, fast data acquisition also calls for dedicated software that enables efficient analysis of the huge 4D image data volumes (e.g., digital image and volume correlation DIC/DVC or 4D particle tracking, see [46–48]). In all these fields, fast progress and innovation is currently ongoing.

4.2.2.5 Experimental 4D Tomography

Fast X-ray tomography opens new possibilities for in-situ and in-operando studies of dynamic processes at pore-scale and even below. For lab-based X-ray tomography, the possibilities of high-speed 3D imaging were reviewed in a recent article by Zwanenburg et al. [49]. With synchrotron-tomography, ultrafast 3D imaging can now be performed at 20 Hz and even faster (Maire et al. [40]). *In-situ mechanical testing* is one prominent *example for the application of fast 4D tomography*. Nano-mechanical devices for compression, tension, and indentation tests, which are specially designed for dynamic tomography investigations, are nowadays commercially available. Also, the preparation of small samples (e.g., pillars with sizes in the range of mm down to a few μm) suitable for nano-mechanical testing is now relatively straightforward due to the availability of focused ion beam (e.g., Xe + plasma FIB) and laser technologies. In-situ tomography during mechanical testing is thus evolving rapidly, for example in the field of alloys [50] and batteries [15].

Fast tomography is also used *in high temperature studies*. Villanova et al. [51] reported the 4D-evolution of microstructures and nucleation of nano-droplets upon sintering of alloys at 700 °C.

So-called in-operando studies enable capturing dynamic processes under real life conditions. *In-operando studies of electrochemical cells* (batteries, fuel cells) are very challenging, because they usually require some in-house development of dedicated experimental equipment, including miniaturization of the cells and of the electrochemical test setup. A tomography system with an efficient optical microscope was designed at the Swiss Light Source (SLS). It was recently used for in-operando studies of *water clusters and two-phase flow in PEM fuel cells* at temporal and spatial resolutions of < 1–10 Hz and < 1–10 μm, respectively [18, 19, 52, 53].

In battery research, synchrotron based in-operando studies have been used to reveal the *4D microstructure evolution upon degradation* and/or (de-)lithiation [54–56].

With laboratory-based systems the acquisition time is generally longer and rather suitable for the characterization of relatively *slow processes in time-lapse mode*. Nevertheless, *two-phase flow in geological samples* was characterized with lab-based tomography at temporal and spatial resolutions of less than 1 min and 15 μm, respectively (Bultreys et al. [27]). The lack of attenuation contrast in two-phase flow can be approached with time-resolved phase contrast imaging as reported by Ohser et al. [57].

4D tomography is thus applicable for dynamic studies in various fields of materials and engineering sciences and also in geology [58–60] as well as in life sciences [33].

The exciting possibilities of modern high-end X-ray tomography open new possibilities for the investigation of tortuosity effects. The exploitation of these opportunities requires dedicated and efficient solutions in image processing, which will be discussed below in Sects. 4.3–4.5.

4.2.3   FIB-SEM Tomography and Serial Sectioning

Commercial dual beam machines combining focused ion beam (FIB) with a scanning electron microscope (SEM) became available around the year 2000. Initially, FIB-SEM was used mainly for failure analysis in semiconductor industries. However, very soon it was recognized that FIB-SEM has a great potential for high-resolution 3D imaging by serial sectioning. First FIB-SEM tomography work was based on in-house developments of a machine-controlled procedure for serial sectioning (Holzer et al. [61]). Already in 2004, voxel resolutions of $6 \times 7 \times 16$ nm could be reached based on the fully automated serial sectioning procedure with integrated drift correction. With a voxel resolution of ca. 10 nm, FIB-SEM tomography opened new possibilities to perform microstructure investigations at the sub-μm scale. FIB-SEM tomography thus became the method of choice for 3D investigations of fine-grained porous media [62–66] at the time when nano-CT was not yet available.

Nowadays, 3D acquisition by 'slice and view' is possible with any commercial FIB-SEM machine. Examples for applications of FIB-SEM tomography cover the fields of geological materials (sandstone, shale, coal) [64, 65, 67–70], zeolite [71], graphite [72, 73], polymers [73], thin films used as optical layers [74], catalysts [75], paper [76] and biomaterials [77]. FIB-SEM tomography is also very important for microstructure investigations of energy materials such as fuel cells [78–88] and batteries [89–94].

FIB-SEM serial sectioning can be used in combination with different detector systems such as EDS (energy dispersive spectroscopy) for mapping element concentrations and EBSD (electron backscattered diffraction) for mapping grain orientations and crystallographic information (see e.g., Uchic et al. [3]). With these analytical detection modes, FIB-SEM tomography became particularly important for the study of metals, alloys, and corrosion science [95–98], but also for battery materials [99–101]. Furthermore, the combination of FIB-SEM tomography with a cryo-transfer system enables the study of delicate, water-containing samples such as cement suspensions, swelling clay and biomaterials [102–105]. Reviews on FIB-SEM tomography and related serial sectioning techniques are given by Holzer and Cantoni [106], Cantoni and Holzer [107], Monteiro and Paciornik [108] and Echlin et al. [109].

As indicated in Fig. 4.2, FIB-SEM tomography initially occupied a niche among other 3D imaging techniques due to its high resolution of ca. 10 nm (nowadays even 5 nm are possible). With X-ray CT, it is only in the last few years that resolutions of less than 100 nm can also be reached. The main limitations of FIB-SEM tomography come from the milling capabilities. With a conventional Ga FIB source, a high milling precision that allows slicing in the range of 10 nm is only possible with relatively low beam currents of ca. 1 nA or less. The corresponding low milling rates lead to relatively long acquisition times of ca. 10–24 h for a stack of 500 to 1000 images. Low milling rates also lead to relatively small sizes of the 3D image window with edge lengths that are typically equal to only a few μm to tens of μm. The milling rates can be increased by using higher ion beam currents, but this comes at the expense of larger beam spot sizes and lower milling precision (i.e., with a decrease of spatial resolution in slicing direction). Fortunately, over the last years, the milling capabilities of FIB-sources improved significantly [110–112] and also new and more efficient serial sectioning techniques were introduced such as Plasma FIB and broad ion beam (BIB) [109, 113]. Hence, as indicated in Fig. 4.2, the various serial sectioning techniques for 3D image acquisition nowadays cover a wide range of voxel resolutions from ca. 5 nm to several μm and a wide range of image window sizes with edge lengths from μm to mm.

4.2.3.1 Basic Principle of Serial Sectioning with a Ga+ FIB-SEM Dual Beam Machine

The FIB-SEM geometry for serial sectioning is illustrated in Fig. 4.6. In a first step, a cube representing the region of interest is exposed with a high beam current for rapid ion milling. The x–y imaging plane, also called 'block-face', is then polished with a lower ion beam current. Subsequently, an SE- or BSE-image is acquired by scanning the block face with the electron beam. A stack of 2D images (i.e., a 3D image volume) is then produced in a fully automated serial sectioning procedure, which consists of two alternating steps: (1) Thin layers of e.g., 10 nm thickness are sequentially removed from the block face with the ion beam and (2) SEM images with a pixel resolution of e.g., 10 nm are acquired from the freshly exposed block face. In this procedure, fiducial markers are used for automated correction of mechanical, magnetic, and electronic drifts. Ideally the milling step size in z-direction is identical to the pixel resolution of the SEM images (x–y plane), which results in isometric voxels.

In most cases, a large number of ca. 1000 or more images would be ideal in order to acquire a representative 3D image volume. The acquisition time for a single slice-and-view cycle includes the following components:

FIB milling time

The milling time depends on the beam current (milling rate), on the size of the imaging plane (x–y) and on the thickness of the milled layer (z-direction).

SEM imaging time

The SEM imaging time depends on the number of pixels (i.e., area of block-face and resolution) and on the dwell time for scanning with the electron beam. Thereby, fast scanning negatively affects the signal-to-noise ratio.

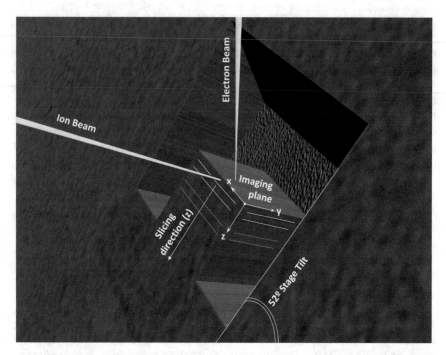

Fig. 4.6 Illustration of serial sectioning with a FIB-SEM dual beam system (taken from Holzer et al. [61])

Time for drift correction(s)

Drift correction is based on images that are taken from specific sample locations, which contain fiducial markers (i.e., reference positions). The time for drift correction(s) thus depends on the imaging conditions (i.e., scan rate, resolution, size of image). In advanced serial sectioning procedures, the drift correction is performed in the x–y plane with SEM- and in the x–z plane with FIB-images.

Time for beam stabilization

The beam requires some time for stabilization after switching from electron to ion beam and back.

The total acquisition time thus depends on various parameters such as the size of the 3D image window, the ion beam-current, the electron scan rate, the detector efficiency. Also contrast (or noise) and sputter rates that are characteristic for the material under investigation have an influence on the acquisition time. Depending on the chosen parameters for serial sectioning, the total acquisition time can thus vary significantly. Typically, in a relatively fast setup with small cubes of a few μm edge lengths, an entire slice-and-view cycle takes ca. 30 s. For a stack with 1000 images, this cycling rate results in a total acquisition time of 8 h and 20 min. The acquisition time can easily increase by a factor of 3 to 4, e.g., for larger cubes (tens of μm), for more precise ion milling and/or slower electron scanning (higher signal-to-noise ratio). Often the number of images in the stack is then reduced to only a few 100 images, in order to shorten the total acquisition time. This leads to 3D image volumes with non-isometric dimensions (e.g., $20 \times 20 \times 5$ μm, whereby the 5 μm direction corresponds to the slicing direction).

4.2.3.2 Trends in Serial Sectioning I: Improvement of Milling Capabilities

Over the last years the milling capabilities for serial sectioning have considerably improved due to the appearance of new ion sources and new milling techniques (for details see Bassim et al. [112] and Echlin et al. [109]):

Conventional FIB

The *liquid metal ion source (LMIS)* is the basis for *conventional Ga FIB*. With beam currents between pA and 100 nA, the sputter rates of Ga FIB machines are relatively low,—especially for organic matter and ceramics. In the meanwhile, *LMIS works with many different metals (Ga, Al, In, Au, Bi) and alloys*. Nevertheless, the milling capabilities of LMIS are still rather limiting. Some improvements of the sputter rates could be achieved with a new 'rocking milling procedure'.

Plasma (P)FIB

Magnetically enhanced inductively coupled plasma FIB sources are capable of significantly higher sputter rates. In addition, at beam currents above 10–50 nA, the *Xe PFIB* provides a much smaller beam diameter than conventional Ga FIB (at similar beam currents). The PFIB thus opens *new possibilities for large area serial sectioning*, whereby the edge length of the image window can reach dimensions in an order of magnitude of 100 μm. Compared to Ga-FIB, the milling with Xe PFIB is ca. 60 times faster, but at the same time PFIB is capable to reveal small slicing thicknesses (and voxel resolutions) of ca. 10 nm. In addition, Xe PFIB produces less beam damage (i.e., the amorphous surface layer is relatively thin) and it is therefore better suited for 3D EBSD compared to conventional FIB. Applications of large area serial sectioning with Xe PFIB are discussed by Burnett et al. [110] and Zhang et al. [111, 114].

Broad Ion Beam (BIB)

The *hollow anode discharge (HAD) Ar source* represents the basis for broad ion beam (BIB) machines, which can be used for sequential milling and polishing of large areas up to the mm²-range. For in-plane milling and polishing with a broad ion beam, *a metal blend is used with a high milling resistance* (e.g., W). HAD Ar ion sources reveal high beam currents up to the μA-range at low beam energies (≤ 5 kV). Milling at low kV induces relatively low beam damage, which makes BIB particularly suitable for 3D EBSD. Generally, the z-resolution (thickness of removed layer) for BIB serial sectioning is in the 100 nm to μm range, but recently more precise BIB-serial sectioning with z-distances as thin as 10 nm were reported [115, 116].

Pulsed laser and combined tri-beam systems

Laser-based systems combined with various microscopy platforms (light microscopy, SEM, FIB-SEM) have been available for many years. Due to limited resolution of the laser, these systems were rather used in the past for targeted feature extraction and micromachining. Thereby, the milling precision of traditional pulsed laser-systems was not suitable for serial sectioning applications. However, modern femtosecond pulsed lasers nowadays provide much higher milling precisions and, at the same time, they cause less beam damage. Recently, a *femtosecond laser* was integrated into a dual beam *PFIB-SEM*, which results in a *tri-beam system*. This device enables precise serial sectioning of large areas in the mm^2 range. With the tri-beam system, the PFIB can be used for fine polishing after efficient milling with the laser. Typically, the step size of laser milling in z-direction is 0.5–1.5 μm [113, 117, 118].

For comparison, the typical performances of the discussed serial sectioning techniques are shown in Table 4.1. This table also includes *ultra-micro tomography serial block face SEM (UMT SBFSEM)* [119], which uses a diamond knife for mechanical sectioning. Furthermore, *robotic serial sectioning by mechanical polishing* [120] is also included for comparison.

Most serial sectioning techniques have a certain tendency towards anisometric voxel resolution. The pixel resolution of the SEM images (i.e., imaging resolution of the x–y plane) is typically in the range of 10 nm. Even large areas up to the mm^2 range that are produced for example with BIB, laser-PFIB tri-beam or ultra-micro tomography can be efficiently scanned at high resolution by using a stitching approach for the SEM imaging. In contrast, for these serial sectioning methods the step size in z-direction is typically limited to ca. 1 μm, which is 20–100 times larger than the SEM pixel resolution (see the column aspect ratio, resolution in Table 4.1).

The dimensions of the 3D image window (i.e., CEL, cube edge lengths of analyzed volumes) also tend to be anisometric. For example, with PFIB and BIB, the total thickness of the image stack (z-direction) that can be acquired at high slicing resolution within reasonable acquisition time is often 10–50 times smaller than the size of the 2D image window in x–y directions (see Table 4.1, column aspect ratio, CEL). These anisometric properties of serial sectioning are also visualized in Fig. 4.2 by elongated ellipses. The long axes of the ellipses indicate different dimensions of voxels and image windows in x–y—(top left part of ellipses) compared to z-directions (bottom right part of ellipses). Large area serial sectioning is thus particularly well suited for the analysis of anisometric samples such as the thin layers of SOFC electrodes (see e.g., Mahbub et al. [121]).

Table 4.1 Comparison of serial sectioning techniques and associated characteristic properties

Sectioning method	Imaging method	2D imaging plane (x,y)			Slicing direction (z)			Aspect ratios (x,y)/z	
		Resolution x,y	No. of pixels x,y	CEL x,y	Resolution z	No. of slices z	CEL z	Resolution	CEL
		nm	–	μm	nm	–	μm	–	–
Ga FIB-SEM	SEM	10	1,000	10	10	500	5	1	2
Xe PFIB	SEM	20	10,000	200	50	1,000	50	0.4	4
Ar BIB	SEM	50	100,000	5,000	1,000	100	100	0.05	50
Laser tribeam	SEM	10	100,000	1,000	1,000	1,000	1,000	0.01	1
UMT SBFSEM	SEM, TEM	10	50,000	500	500	100	50	0.02	10
mech. polishing	Light micr.	500	20,000	10,000	100,000	100	10,000	0.005	1

For most serial sectioning techniques, the resolutions tend to be different for the x–y imaging plane and for the sectioning direction (z). Furthermore, the number of acquired images (slices in z-direction) is usually relatively small due to long acquisition times. This results in anisometric data cubes with aspect ratios different from 1. (CEL = cube edge lengths of a 3D image window). Note that the properties (e.g., resolution, number of pixels) are estimations of characteristic values. In reality, these values can vary significantly, depending on the chosen acquisition parameters

4.2.3.3   Trends in Serial Sectioning II: Imaging Capabilities and Detection Modes

A significant advantage of destructive serial sectioning compared to X-ray tomography comes from the fact that the exposed surfaces (block-faces) can be probed with many different imaging and surface characterization techniques. Thereby, SEM based serial sectioning benefits considerably from the progress in low-voltage SE- and BSE-imaging, which provide high contrast at high resolution. This progress is mainly due to the innovative improvement of in-lens or through-the-lens detectors [107]. In addition, fast spectral and elemental mappings with silicon drift EDS detectors open-up new possibilities in 3D chemical mapping. Furthermore, new EBSD cameras enable grain orientation mappings with significantly shorter acquisition time, higher spatial resolution, and larger image window size. As discussed by Echlin et al. [109], there is a clear trend in serial sectioning tomography towards larger size of the image windows (e.g., with PFIB, Laser-tribeam or BIB) due to the possibility of combining high milling rates with a high resolution. Another important trend is the evolution towards simultaneous acquisition of multiple signals, which is also called multi-modal tomography (i.e., serial sectioning with simultaneous acquisition of EDX or EBSD together with SE, BSE and even with SIMS), see e.g., the 3D FIB EBSD image data considered in [99, 100].

4.2.4   Electron Tomography

Transmission electron microscopy (TEM) enables for microstructure analysis at the nanoscale and even with atomic resolutions. Due to the invention of aberration corrected lenses, probe sizes as small as 0.05 nm can be reached with TEM [122]. In electron tomography (ET) numerous TEM projections are acquired in a tilt series at different angles, from which the corresponding 3D structure can be reconstructed. Current trends in nano-tomography (both, in ET and X-CT) were recently reviewed by Yan et al. [34].

Distinct ET methods have been developed separately for physical and biological sciences in order to overcome the specific sample-based limitations [123]. In materials science, ET is particularly important for the study of functional materials such as nano-porous materials for chemical engineering, nanoparticle agglomerations or nanostructured catalysts in fuel cells [124, 125].

A major strength of ET is obviously its high resolution-power. However, a relatively short mean free path length of electrons puts strong limitations to the maximum sample thickness, which is ca. 100 nm for mid Z-materials at 200 keV. In nano-tomography mode, ET is typically performed with a resolution of ca. 0.5 nm. The maximum sample and image window sizes are then typically not more than 100–300 nm. In atomic scale tomography mode, ET is performed with < 0.1 nm resolution. The corresponding image window size is then typically not more than 10–20 nm.

Moreover, a particular strength of ET is the ability to detect different signals from the same sample. Thereby one has to distinguish between full field transmission (TEM) and scanning transmission modes (STEM). Full field imaging allows for faster acquisition and low dose imaging of delicate samples. High Z-contrast is achieved in STEM with dark field (HAADF) and bright field (BF) modes. STEM also enables spectroscopic tomography whereby chemical maps are collected with EDS (energy dispersive spectroscopy) or EELS (electron energy loss spectroscopy). In addition, new detection modes are currently evolving, which provide interesting information about the spatial distribution of magnetic and electric fields, strain, grain orientation and/or crystallographic defects (see e.g., [34]).

Current improvements aim to push the limits of ET in various directions:

Acquisition time

The acquisition time is typically in the range of several hours, due to time consuming tilt by tilt tracking of objects. In future, automated repositioning can shorten the acquisition time considerably.

3D reconstruction

The precision of the 3D reconstructions is limited due to a relatively small tilt range (missing wedge problem) and due to a relatively low number of projections. New algorithms based on machine learning are capable to reveal much better 3D reconstructions, despite these limitations.

Sample holders and sample fabrication

New sample holders and stages, as well as improved sample fabrication procedures with automated FIB (producing cylindrical instead of lamellar samples) will contribute to better data acquisition and more reliable 3D reconstructions.

Detector technology

Important improvements can also be expected with respect to the detectors, which are capable to capture different signals (as mentioned above) with higher sensitivity, better signal-to-noise ratio, and faster acquisition time.

4.2.5   Atom Probe Tomography

Atom probe tomography (APT) is capable to perform 3D analysis at the atomic scale (around 0.1–0.3 nm resolution in depth and 0.3–0.5 nm laterally). Electrochemical polishing and focused ion beam (FIB) methods are used for sample preparation in the form of a very sharp tip. A very high electrostatic field (in an order of magnitude of 10 V/nm) is induced at the sharp tip, which is slightly below the point of atom evaporation. Laser or HV pulsing is then superimposed, in order to evaporate single atoms from the tip surface by a field effect (near 100% ionization). The atoms or ions are collected very efficiently with a position sensitive detector (PSD). The detector allows measuring simultaneously the mass of the ions (more precisely: the mass-over-charge ratio) by time of flight and at the same time to reconstruct the original position of the atom on the tip surface. The atoms are progressively removed from the tip so that a 3D image of the material can be reconstructed at the atomic scale.

APT has been successfully applied in materials science for many years, in particular for metals, alloys, and semiconductors (e.g., for the study of interfaces and inter-diffusion phenomena). A review on APT investigations of aluminium alloys was recently given by Ceguerra and Marceau, 2019 [126]. Technical advancements such as the introduction of pulsed laser-assisted field evaporation also enable atom probe analysis of oxides, which extends the field of APT applications to geological materials and metal corrosion (see e.g., Eder et al. [127]).

Air- and temperature-sensitive samples require transfer systems between FIB and atom probe under both vacuum and cryogenic conditions [128, 129]. Such a cryo-transfer system was recently used to study corrosion of nuclear glass. The sample consisted of a nano-porous gel filled with liquid electrolyte. It was shown for the first time that APT is capable to describe the 3D distribution of chemical concentrations at solid–liquid interfaces with (near) atomic resolution (Perea et al. [130]). The size of the 3D image window was $20 \times 20 \times 20 \, nm^3$. APT thus enables to detect variations in the chemical composition of the electrolyte and to combine this chemical information with structural information of tortuous pathways in the nano-porous network.

4.2.6 Correlative Tomography

For the investigation of complex microstructures, the application of a single microscopy method with a fixed resolution and/or with a single detection mode is sometimes not suitable for a representative characterization. For example, in materials with a wide pore size distribution, nano-tomography may be capable to capture small pores and bottlenecks, but the image window is then often too small for capturing the larger pores in a representative way. Alternatively, low-resolution tomography that provides a larger and representative image window may not be capable to resolve the smaller pores and bottlenecks. Fortunately, the contradictory requirements of a high resolution and a large representative image volume can be satisfied with the help of correlative tomography, which makes use of two or more tomography methods with different resolutions and image window sizes. Furthermore, in multi-phase materials, correlative tomography can also be used to capture multimodal information. The combined detection of Z-contrast, chemical- and crystallographic information can then be used as a basis for reliable interpretation, segmentation, and phase identification.

The power of correlative microscopy for advanced microstructure characterization has been recognized for many years. Thereby, complementary microscopy methods with different resolutions and detection modes are applied for the same regions of interest (RoI). Image registration can then be used to combine the information of the spatially overlapping data sets [131, 132]. Initially correlative microscopy was mainly based on the combination of 2D microscopy methods such as light and fluorescence microscopy, AFM, SEM, (S)TEM, (S)XTM (e.g., [133, 134]). However, very soon correlative imaging approaches were also combining 2D microscopy with tomography (see e.g., Caplan et al. [135]). Nowadays, due to the progress in 3D imaging and 3D image processing, the number of studies applying correlative tomography is rapidly increasing. Correlative tomography enables characterizing the full complexity of disordered microstructures by combining multi-modal, multi-scale and multi-dimensional information acquired with multiple 3D techniques from the same region of interest (or from overlapping RoIs).

A full review of correlative tomography is beyond the scope of this article. Overviews of correlative tomography are given by Burnett and Withers [136, 137] for materials science applications, as well as by Bradley and Withers et al. [138] for biomaterials.

In correlative tomography various combinations of 3D techniques are possible. Typically, non-destructive methods at lower resolution such as X-ray CT or confocal laser scanning microscopy are used in a first step. Subsequently, destructive 3D methods (e.g., 3D FIB-SEM, APT, ET) in combination with site-specific sampling techniques (e.g., with laser and with FIB lift-out techniques) are used for zoom-in characterization at higher resolutions. To illustrate the evolution of correlative tomography we briefly present some literature examples from the last 10 years, which are also summarized in Table 4.2.

In Caplan et al. [135], the correlation of various 2D and 3D methods are discussed in context with a thorough characterization of biomaterials.

Tariq et al. [11] used multi-scale tomography (XCT and FIB-SEM) for the characterization of hierarchical pore structures in ceramics. The cumulative pore size distributions (PSD) obtained with multi-scale tomography are different from those obtained with mercury intrusion porosimetry (MIP). The example illustrates that it is difficult to quantify hierarchical pore structures based on information from different methods (experimental vs. imaging) and different length scales. New up-scaling approaches are needed for integration of multi-scale information in hierarchical pore networks.

Shearing et al. [139] investigated the microstructure of lithium-ion battery electrodes with XCT at different length scales. It was possible to obtain consistent results for porosity, tortuosity, and surface area with different CT scans. Apparently, with the chosen resolutions and sizes of data volumes, it was possible with different tomography methods to capture the relevant features in a representative way.

Burnett et al. [136] used correlative microscopy for the study of metal corrosion, combining multi-scale tomography with 2D maps from EBSD and EDS. This approach enabled to distinguish between pitting and inter-granular corrosion phenomena.

Bradley and Withers [138] used correlative tomography for characterization of biological materials with hierarchical microstructures and anisotropic mechanical properties.

Table 4.2 Examples of correlative tomography studies from the last 10 years, illustrating the methodological evolution and trends

1st Author		Caplan et al.	Tariq et al.	Shearing et al.	Burnett et al.	Bradley + Withers	Saif et al.	Kwia-towski	Fam et al.	Keller et al.	Meyer et al.
Year		2011	2011	2012	2015	2016	2017	2017	2018	2018,2013	2019
Reference		[135]	[11]	[139]	[136]	[138]	[70]	[140]	[141]	[142, 143]	[16]
Material		Biomat	Ceramics hierarch. pores	LIB electrodes	Metal corrosion	Biomat	Oil shale pyrolysis	Fe–Mn-steels	Au catalyst hierarch. pores	(Opalinus) Clay	PEM fuel cell
1st 3D method		ET	syncCT	XCT	XCT	XCT	XCT	APT	ptych. XCT	sync XCT	XCT
Voxel size	$(\text{x nm})^3$		1,400	600	3,800	Wide range	5,000	0.1	13	440	372
Volume	$(\text{x μm})^3$		1,500	220	4,000	of lengths scales …	10,000	0.1	10	440	700
2nd 3D method		CLSM	FIB-SEM	XCT	XCT	… including	XCT	ET	FIB-SEM	FIB-SEM	XCT
Voxel size	$(\text{x nm})^3$		43	65	800	time lapse	800	0.1	13	15	63
Volume	$(\text{x μm})^3$		20	45	400	tomography	1,000	0.1	10 x 12 x 3	4.5	5-65
3rd 3D method		FRET		sync XCT	FIB-SEM		FIB-SEM		ET	STEM	
Voxel size	$(\text{x nm})^3$			15	8 x 10 x 50		14 x 18 x 10		1.3	0.5–1	
Volume	$(\text{x μm})^3$			7	< 100		28 x 20 x 6		3 x 3 x 0.3	0.1–0.3	
4th 3D method		FIB-SEM									
1st 2D method		LM			EBSD		HR SEM	STEM		EDX	He-FIB
Pixel size	$(\text{x nm})^2$				50			0.05		1	0.5–60
Image window	$(\text{x μm})^2$				15			< 0.1		0.1	0.15–100
2nd 2D method		(S)TEM			TEM EDX		SEM EDS				TEM
Pixel size	$(\text{x nm})^2$				50						0.05–0.2
Image window	$(\text{x μm})^2$				15						0.08–0.3
3rd 2D method		AM					MAPS				
4th 2D method		SEM									
Bulk experimental		MIP	MIP	XRD							

Saif et al. [70] applied multi-scale tomography in combination with various 2D methods (MAPS, high resolution SEM, stitching of multiple SEM images) for characterization of oil shale pyrolysis. The multi-scale and multi-modal information enabled a thorough characterization of the heterogeneous clay microstructures, including accurate identification of porosity, organic matter, and mineralogical composition.

Kwiatowski da Silva et al. [140] used correlative TEM (ET) and atom probe tomography (APT) in combination with multi-scale modeling for characterization of Fe–Mn steels. This approach provides unique insight on the mechanism of Mn segregation to edge dislocations.

Fam et al. [141] used several tomography methods (nano-CT, FIB-SEM, ET) at high resolutions (1–15 nm) for the characterization of hierarchical structures in nano-porous gold catalysts. The results for porosity and pore size vary depending on the method, even though the resolutions were not very different. Most probably this puzzling picture arises from different contrast modes, which have a strong impact on the phase segmentations and associated quantitative analyses.

Keller and Holzer [142] and Keller et al. [143] used XCT, FIB-SEM and ET for a thorough characterization of pores in Opalinus clay. A concept for image-based up-scaling from micro- to meso-scale porosity and associated estimation of permeability is presented. This approach is also capable of capturing the anisotropic transport properties of clays across lengths scales from nm to mm.

In a recent study on PEM Fuel cells, Meyer et al. [16] combined multi-scale XCT with high resolution 2D imaging by He-FIB and TEM. The different methods give complementary information, which is important for accurate identification of relevant features in the heterogeneous multi-layer assembly, such as Pt nanoparticles in the micro-porous catalyst layer (MPL) and meso-pores in the gas diffusion layer (GDL).

For some of the mentioned correlative studies, the achieved resolutions and image window sizes of the 3D datasets are plotted in Fig. 4.7. This Figure illustrates that data volumes produced in current correlative tomography studies are usually smaller than $1,000^3$ voxels. This is particularly true for the nano-tomography methods. The use of relatively small data volumes in correlative tomography contrasts the general trend of 'non-correlative' tomography (i.e., using only one single tomography method), whereby the limits are pushed towards larger image windows and larger data volumes (e.g., $10,000^3$ voxels). This comparison points to a certain potential for future development of correlative tomography towards larger image windows, which is particularly helpful for the characterization of materials with complex, heterogeneous microstructures.

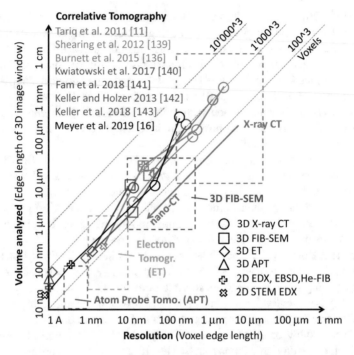

Fig. 4.7 Plot of 3D image datasets from correlative tomography studies (see Table 4.2). It illustrates that 3D datasets from correlative tomography are similar or even smaller smaller than $1,000^3$ voxels, which is opposite to the current trend in 'normal' tomography (XCT, PFIB-SEM etc.), where the limits are pushed towards larger image windows and larger data volumes (e.g., $10,000^3$ voxels)

4.3   Available Software Packages for 3D Image Processing and Computation of Tortuosity

As discussed in Chap. 2, there exist numerous types of tortuosities. In the following sections it is discussed how these different tortuosities can be extracted from 3D images. In our description we use the classification scheme and tortuosity nomenclature that was introduced in Chap. 2 (see Fig. 2.8).

The processing and analysis of 3D image data is a complex task. Fortunately, there are numerous software (SW) packages available nowadays for 3D image processing and also for pore scale modeling. Table 4.3 represents a list of these software (SW) packages. The SW packages offer a wide range of applications and opportunities, which are typically structured in different modules. In the following section we will discuss the capabilities of the SW packages with their different modules.

Table 4.3 Compilation of available SW packages with various modules for 3D image processing, modeling, and simulation

No.	Software name	Proprietary C, FW, NA Commercial, FreeWare, not avail.	IP I: (qualitative)	Microstr. modeling Stochastic or DEM	Image processing IP II: (quantitative analysis) various characteristics	Tortuosity dir geom	mixed	indir phys	Numerical simulation V voxel M mesh	Type of transport cond/diff (Laplace)	flow (Navier-Stokes)	Multiphys e.g. electro-chem, CFD, mechanical	Numerical method(s)
1) Multi-modular SW for 3D microstructure analysis and modeling													
1	GeoDict	C	x	x	x	x	x	x	V	x	x		FVM, RWM
2	Digimat	C	x	x				x	M	x	x	x	FEM
3	PuMA porous Microstr. Analysis	FW	x	x	x			x	V	x			FDM, RWM
4	MICRESS (for metals and alloys)	FW	x	x	x				M			x	MultiPhaseFlow CFD
5	Avizo / Amira	C	x		x	x		x	V	x	x		FVM
6	PerGeos for Digital rock analysis	C	x		x	x		x	V	x	x		FVM
7	Simpleware (ScanIP)	C	x		x	x		x	M	x	x	x	FEM
8	VGS Studio Max	C	x		x			x	M	x	x	x	FEM
9	Pore3D	FW	x		x	x		x	V	x	x		FVM
2) SW for tortuosity analysis													
10	TauFactor	FW			x			x	V	x			Matlab, FDM
11	Tort3D	NA				x							Matlab
12	BruggemannEstimator (Batteries)	FW		x (2D input)				x					Mathematica
13	Fiji XLib	FW	x		x								Fiji-plugin

(continued)

Table 4.3 (continued)

No.	Software name	Proprietary C, FW, NA Commercial, FreeWare, not avail.	IP I: (qualitative)	Microstr. modeling Stochastic or DEM	various characteristics	Tortuosity dir geom	Tortuosity mixed	Tortuosity indir phys	V voxel M mesh	cond/diff (Laplace)	flow (Navier-Stokes)	Multiphys e.g. electro-chem, CFD, mechanical	Numerical method(s)
14	Fiji skeletonize3D, AnalyzeSkeleton	FW			x	x							Fiji-plugin
15	Pytrax	FW					x		V	x			RWM
16	Mixed tortuosity (Vav) see Matyka + Koza, 2012 [152] (implementation)						x		V	x	x	x	FVM, FDM, LBM
	3) SW for 3D image processing and visualization (data import, 3D visualization, filtering, segmentation)												
17	DragonFly	C	x		x	x							
18	Matlab	C	x		x	x		x				x	FEM-toolbox
19	Mathematica	C	x		x								
20	ImageJ, Fiji	FW	x		x	x							
21	ADCIS / Aphelion / SDK	C	x										
22	Imaris / Bitplane	C	x										
23	Huygens SW	C	x										

(continued)

Table 4.3 (continued)

No.	Software name	Proprietary C, FW, NA Commercial, FreeWare, not avail.	IP I: (qualitative)	Microstr. modeling Stochastic or DEM	Image processing IP II: (quantitative analysis) various characteristics	Tortuosity dir geom	mixed	indir phys	Numerical simulation V voxel M mesh	Type of transport cond/diff (Laplace)	flow (Navier-Stokes)	Multiphys e.g. electro-chem, CFD, mechanical	Numerical method(s)
24	MeVisLab	C	x										
25	Itk: Insight Toolkit	FW	x										
26	Itk-SNAP	FW	x										
27	SimpleITK	FW	x										
28	VTK: Visualization toolkit	FW	x										
29	TTK: Topology toolkit	FW	x										
30	ParaView	FW	x										
31	Mango	FW	x										
32	SciKit_Image	FW	x										
33	3D-Slicer	FW	x										
34	vaa3d	FW	x										
35	VisIt	FW	x										
36	VisNow	FW	x										
37	TomoWarp2	FW	x										DVC

(continued)

Table 4.3 (continued)

No.	Software name	Proprietary C, FW, NA Commercial, FreeWare, not avail.	IP I: (qualitative)	Microstr. modeling Stochastic or DEM	Image processing IP II: (quantitative analysis) various characteristics	Tortuosity dir geom	mixed	indir phys	Numerical simulation V voxel M mesh	Type of transport cond/diff (Laplace)	flow (Navier-Stokes)	Multiphys e.g. electro-chem, CFD, mechanical	Numerical method(s)
38	ImageVis3D	FW	x										
39	Morpho +	NA	x		x								
40	StrainMaster	C	x		x								DIC, DVC
41	MIST	FW	x		x	x							Matlab

4) SW for tomography data (Image processing solutions such as 3D reconstruction specially designed for a certain type of tomography)

No.	Software name	Proprietary	IP I:										
42	efx-ct: SW from X-CT manufacturer	C	x										
43	TomoPy, for (synchrotron) X-CT	FW	x										
44	ASTRA Toolbox, for (synch.) X-CT	FW	x										
45	Tomviz: ET Electron Tomo (S)TEM	FW	x										
46	Protomo: for ET, (S)TEM	FW	x										

(continued)

Table 4.3 (continued)

No.	Software name	Proprietary C, FW, NA Commercial, FreeWare, not avail.	IP I: (qualitative)	Microstr. modeling Stochastic or DEM	Image processing IP II: (quantitative analysis) various characteristics	Tortuosity dir geom	mixed	indir phys	Numerical simulation V voxel M mesh	Type of transport cond/diff (Laplace)	flow (Navier-Stokes)	Multiphys e.g. electro-chem, CFD, mechanical	Numerical method(s)
47	Pytom: for Cryo ET, (S)TEM	FW	x										
48	TomoJ, ImageJ Plugin for ET	FW	x										
49	Atomprobelab: for APT	FW	x										
50	threedepict for APT	FW	x										
51	apttools for APT	FW	x										
52	Dream3D: for EBSD from FIB-SEM	FW	x	x	x								
	5) SW for numerical modelling (multi-physics simulation: CFD, electrochemistry, structural mechanics—at micro- to macro-scales)												
53	Comsol	C							M	x	x	x	FEM
54	Simcenter Star CCM +	C							M	x	x	x	FVM, DEM
55	Ansys suite, Fluent	C							M	x	x	x	FVM, VOF
56	Flow3D	C							M	x	x	x	CFD, VOF
57	Abacus FEA, Simulia	C							M		x	x	FEM

(continued)

Table 4.3 (continued)

No.	Software name	Proprietary C, FW, NA Commercial, FreeWare, not avail.	IP I: (qualitative)	Microstr. modeling Stochastic or DEM	Image processing IP II: (quantitative analysis) various characteristics	Tortuosity dir geom	mixed	indir phys	Numerical simulation V voxel M mesh	Type of transport cond/diff (Laplace)	flow (Navier-Stokes)	Multiphys e.g. electro-chem, CFD, mechanical	Numerical method(s)
58	Advanced Sim. Library ASL	C							(rect. grid) M	x	x	x	FDM LBM levelset
60	ADINA	C							M	x	x	x	FEM
61	FEATool Multiphysics	C							M	x	x	x	FEM
62	SIMSCALE	C							M	x	x	x	LBM, FEM, CFD
63	SRM	C											
64	OpenFoam	FW							M	x	x	x	CFD, FVM, VOF
65	Elmer	FW							M	x	x	x	FEM
66	FreeFEM	FW							M	x	x	x	FEM
67	Su2	FW							M	x	x	x	CFD
68	SESES	FW							M	x	x	x	FEM
69	FenICs	FW							M	x	x	x	FEM
70	Salome	FW	x										Pre-, Postproc
71	OpenLB	FW						x	V	x	x	x	LBM

(continued)

Table 4.3 (continued)

No.	Software name	Proprietary C, FW, NA Commercial, FreeWare, not avail.	IP I: (qualitative)	Microstr. modeling Stochastic or DEM	Image processing IP II: (quantitative analysis) various characteristics	Tortuosity dir geom	mixed	indir phys	Numerical simulation V voxel M mesh	Type of transport cond/diff (Laplace)	flow (Navier-Stokes)	Multiphys e.g. electro-chem, CFD, mechanical	Numerical method(s)
72	Palabos (Parallel LB solver)	FW						x	V	x	x	x	LBM
6) SW for 3D microstructure modeling (creation of virtual 3D microstructures by stochastic modeling or DEM)													
73	ESyS	FW		x									
74	GenGeo	FW		x									
75	YADE	FW		x									
76	Mote3D	FW		x									
77	PFC Particle Flow Code	C		x									

An extended version with the corresponding web-links and further details can be downloaded from the electronic appendix (supplementary file 4.1). *Legend* IP = image processing, DEM = discrete element method, FEM/FVM/FDM = finite element/volume/difference method, RWM = random walk method, LBM = lattice Boltzmann method, CFD = computational fluid dynamics, VOF = volume of fluid, DIC/DVC = digital image/volume correlation

4.3.1   Methodological Modules

The different columns in Table 4.3 from left to right represent specific methodological modules, which are used in the workflows for the characterization of different tortuosity types (see the workflow in Fig. 4.1).

4.3.1.1   Image Processing I (Qualitative)

Modules for qualitative image processing (IP) provide solutions for 3D reconstruction, filtering of image defects, segmentation, and visualization. Some SW packages also provide an IP module for mesh generation. The meshed 3D data is often used as a basis for numerical simulations, e.g., with finite element models.

4.3.1.2   Stochastic Microstructure Modeling

Such SW modules enable the generation of virtual 3D microstructures with stochastic modeling or with discrete element modeling (DEM), which is an important option for data driven, statistical investigations of micro–macro relationships (see Chap. 5).

4.3.1.3   Image Processing II (Quantitative)

Modules for quantitative image processing are used for the determination of morphological microstructure characteristics. In particular, dedicated SW modules are used for characterization of direct geometric tortuosities ($\tau_{dir\_geodesic}$, $\tau_{dir\_medial\_axis}$, $\tau_{dir\_skeleton}$, τ_{dir_PTM}, $\tau_{dir\_percolation}$, τ_{dir_FMM}, $\tau_{dir\_pore\_centroid}$) and also for mixed tortuosities.

It must be emphasized that the determination of indirect-tortuosities ($\tau_{indir\_phys\_sim}$) and mixed tortuosities (τ_{mixed_phys}) cannot be performed with geometric image analysis alone, because the determination of these tortuosities requires modules for 3D numerical simulation of the underlying transport process and extraction of effective or relative transport properties.

In addition to tortuosity, several other morphological characteristics are important in the context with effective transport properties, which are the following:

- solid and pore volume fractions (ϕ, ε)
- continuous pore (or solid phase) size distributions ($cPSD$)
- mean radius of pore bulges ($r_{max} = r_{50\_cPSD}$)
- simulated mercury intrusion porosimetry ($MIP\text{-}PSD$)
- mean bottleneck radius ($r_{min} = r_{50\_MIP\text{-}PSD}$)
- constrictivity (β)
- hydraulic radius (r_h)

- specific surface and interface areas (*SSA, SIA*) and surface areas per volume, respectively (S_V, I_V)

Further morphological microstructure descriptors for microstructure characterization can be found in [144–146].

4.3.1.4 Numerical Simulation of Transport/Pore Scale Modeling

We distinguish *two main groups of SW packages* with modules for 3D-simulation of physical and/or chemical (transport) processes:

Voxel-based simulations

SW packages that are powerful in micro-scale simulations use voxel-based images to capture the structural input and they typically solve one specific transport equation at a time (e.g., Navier–Stokes solver for viscous flow).

Mesh-based simulations

SW-packages that are strong in solving coupled processes (e.g., by coupling transport with electrochemistry, with thermal behavior and/or with mechanics) typically use a mesh-based representation of the structural input. The mesh-based representation reduces the data volume significantly. But this benefit comes at the cost of lesser morphological details and precision.

SW packages for multi-physics simulations operating with mesh-based input are thus rather suitable for macro-homogeneous modeling, whereas SW packages operating with voxel-based input are better suited for microstructure simulations.

The SW *tools for pore scale modeling* are of particular importance for the *computation of the two following tortuosity categories*:

Indirect physics-based tortuosities

Indirect tortuosities (τ_{indir_phys} with variations τ_{indir_ele}, τ_{indir_therm}, τ_{indir_diff}, τ_{indir_Kn}, τ_{indir_hydr}) are computed from effective transport properties (conductivity σ_{ele}, σ_{therm}, diffusivity D_{eff}, D_{Kn} or permeability κ), which can be obtained from numerical transport simulations.

Mixed tortuosities

Mixed tortuosities ($\tau_{mixed\_phys\_streamline}$, $\tau_{mixed\_phys\_Vav}$, with *phys = ele, therm, diff* or *hydr*) are computed by image analysis from 3D vector fields, which represent the local flux within the pore structure. SW packages that enable to calculate the mixed tortuosities must be capable of performing both, numerical transport simulations (preferably voxel-based) and quantitative image analysis of 3D vector fields.

4.3.2   Different Types of SW Packages

The available SW packages can be grouped according to the modules that they include. The following six groups are distinguished in Table 4.3 (*from top to bottom*):

4.3.2.1   Multi-modular SW Packages

Multi-modular SW packages for 3D microstructure analysis and microstructure modeling provide combined solutions for image processing, quantitative microstructure analysis and numerical simulation. An important characteristic is the capability to perform transport simulations with voxel-based structural input. This option is available, for example, in the SW packages GeoDict (Math2Market), PuMa (NASA), Avizo/Amira (ThermoScientific), PerGeos (ThermoScientific) and Pore3D (Elettra Scientific). Voxel-based simulations are capable to capture the microstructure input from tomography with higher precision and accuracy compared to meshed-based simulations. Another important characteristic is the ability to determine and compare different types of tortuosities based on quantitative image processing and numerical modeling. In this context, *GeoDict is currently the only SW package that enables characterizing all three tortuosity categories (direct geometric, indirect physics-based and mixed tortuosities)* via *the included Compute Tortuosity App.* The third important characteristic refers to the option of stochastic modeling, which is used to generate virtual 3D microstructures, so-called digital twins. This option is provided, e.g., by GeoDict, Digimat (eXtreme engineering), PuMA and Micress (RWTH-Aachen). These multi-modular SW packages also provide the exciting opportunities to perform digital materials design (DMD). Thereby numerous 3D microstructures can be created, and their performances can be characterized by virtual testing (using voxel-based numerical simulations). Based on the combination of stochastic microstructure modeling, virtual testing and quantitative image analysis, these SW packages also provide outstanding opportunities for statistical investigations of microstructure-property relationships (see Sect. 4.7 and Chap. 5).

4.3.2.2   SW Packages for Tortuosity Analysis (Quantitative Image Processing and Numerical Simulation)

Some SW packages are specifically developed for tortuosity analysis (IP II). A prominent example is *TauFactor from Imperial College London* (Cooper et al. [147]), which is a Matlab code for voxel-based simulations of diffusive transport using the finite difference method. It provides physics-based indirect tortuosity (τ_{indir_diff}) and the associated tortuosity factor (see Chap. 2, Eq. 2.15: $T = \tau^2$), respectively. Furthermore, TauFactor is also capable to compute various other microstructure characteristics such as porosity, surface area and three phase boundary (TPB) length.

The *BruggemanEstimator* is a Mathematica code developed *at ETH Zurich* (Ebner and Wood [148]), which uses the Bruggeman relation (see Chap. 3, Eq. 3.3: $\tau = \varepsilon^{\alpha}$) for estimation of indirect tortuosity in granular materials. It is primarily designed for the characterization of battery electrodes. Two orthogonal 2D images are used as input for the statistical analysis of particle shapes and particle orientations. Differential effective medium theory is then applied as a tool to predict the Bruggeman exponent (α) and the associated indirect tortuosity.

Fiji plugins for skeletonization (imagej.net/Fiji: skeletonize3D, AnalyzeSkeleton) can be used for the determination of geometric tortuosity ($\tau_{dir\_skeleton}$). Moreover, various other microstructure characteristics such as *cPSD*, *MIP-PSD* and constrictivity can be determined with the XLib plugin in Fiji (imagej.net/Fiji: XLib [66, 149]).

The SW package *MIST* [150] for image processing also provides tools for the computation of the geometric tortuosity defined in [151].

Dedicated SW for the computation of mixed tortuosities is rare. Matyka and Koza [152] describe how to implement an in-house code for analysis of volume-averaged tortuosity ($\tau_{mixed\_phys\_Vav}$).

For completion it must be emphasized that many of the SW packages in 4.3.2.1 also provide interesting options for tortuosity characterization.

4.3.2.3   SW Packages for Qualitative 3D Image Processing and Visualization

Numerous SW packages are available for qualitative image processing (IP I) and visualization. They provide various options for raw data import from tomography, 3D reconstruction and visualization, filtering of noise and correction of image defects (e.g., background correction), segmentation and mesh generation. Some of these image-processing modules are embedded within a larger commercial SW package (e.g., image-processing toolkit in Matlab and in Mathematica). ImageJ/Fiji [153] is an important freeware for image processing. Moreover, numerous SW packages for image processing are developed for medical and life science applications (e.g., ITK, VTK, TTK etc.). However, in many cases, the life science-oriented SW packages rarely consider morphological characteristics that are frequently used in physical and engineering sciences (tortuosity, constrictivity, pore size distributions).

4.3.2.4   SW Packages for Specific Tomography Data

Some SW packages provide dedicated image-processing solutions such as 3D reconstruction, which are specially designed for a certain type of tomography (i.e., processing of raw data from a specific tomography methodology). For example, the ASTRA toolbox [154] is designed for processing of raw data from synchrotron Xray CT. TomViz is dedicated to data processing from electron tomography (TEM, STEM), and Dream3D for processing and analysis of EBSD data from FIB-SEM tomography.

4.3.2.5   SW Packages for Numerical (Multi-physics) Modeling

Numerous SW packages are available for so-called multi-physics simulations. They are usually based either on finite element (FEM), finite difference (FDM), finite volume (FVM) or lattice Boltzmann methods (LBM). The SW packages for numerical modeling are particularly strong in simulating coupled processes (i.e., combinations of CFD, transport, electrochemistry, structural mechanics etc.) at different lengths-scales. Well-known commercial SW packages of this type are for example Comsol, Simcenter Star CCM + (Siemens), Ansys/Fluent and Abacus (Dassault). There are also powerful freeware packages and libraries available like OpenFOAM, FreeFEM, FEniCS or SESES.

In most cases the multi-physics modeling approach makes use of a mesh-based structural input. However, as mentioned above, precise descriptions of complex 3D microstructure information from tomography are difficult to achieve with mesh-based representations. Therefore, most SW-packages for multi-physics simulation are better suited for simulations at macro-homogeneous scales and/or scenarios with relatively simple morphologies. Mesh-based simulations are thus not recommended for tortuosity analysis of complex microstructures. Note that GeoDict offers packages for voxel-based multi-physics simulation on the microstructure scale for specific applications (i.e., electrochemistry, structural mechanics, digital rock physics and filtration).

4.3.2.6   SW Packages for 3D Microstructure Modeling

Besides the examples mentioned in part a), only a few additional SW packages are available, which can be used for the generation of virtual 3D microstructures. Freeware packages like ESyS, GenGeo, Yade and Mote3D are based on purely geometric packing of particles using the discrete element method (DEM).

The particle flow code (PFC, Itasca Consulting Group) enables virtual particle packing based on physical interactions (i.e., simulating mechanical densification and/or particle growth and crystallization). As mentioned above, also some multi-modular SW packages (e.g., GeoDict and PuMa) offer the option to generate virtual 3D microstructures. In particular, GeoDict offers specific modules for virtual design of granular and fibrous 3D microstructures. A short review of methods and models from stochastic geometry for the creation of virtual 3D microstructures is given in Sect. 4.7.

In the following Sects. (4.4–4.7), the workflow from 3D image acquisition to quantitative analysis of tortuosity is discussed in more detail.

4.4   From Tomography Raw Data to Segmented 3D Microstructures: Step by Step Example of Qualitative Image Processing

After image acquisition with a suitable tomography method, it is necessary to transform the raw data into a segmented 3D microstructure (see workflow in Fig. 4.1). This transformation typically includes the following steps of qualitative image processing (IP I):

- corrections of image defects (e.g., noise filtering, background removal and contrast leveling),
- 3D reconstruction (e.g., alignment of FIB-stack or filtered back projection of CT scans) and, finally,
- segmentation (i.e., phase identification, object recognition, labeling).

Image processing procedures for all these steps are well established and suitable algorithms are implemented either in freeware or in commercial software packages (see Table 4.3, column 'Image Processing I'). Details on qualitative image processing can also be found in textbooks and review articles (e.g., Russ [155], Schlüter et al. [156]). Note that the use of machine learning leads to significant advances in the field of image processing. For machine learning algorithms, which are powerful for image segmentation, we refer, e.g., to [157, 158]. Examples of hybrid approaches combining classical image analysis with machine learning are presented in [100, 159], while a deep neural network is trained in [160], which allows for a reliable segmentation of FIB-SEM image data even if shine-through artifacts are present.

Nevertheless, it must be emphasized that it is very difficult to establish standardized procedures for 3D reconstruction and segmentation that allow user independent automation, because each raw data set is somehow unique due to the specific underlying settings associated with the tomography method, the used imaging parameters and the specific sample and materials properties. For each dataset, a careful adaptation of the image processing procedure for 3D reconstruction and segmentation is thus very important in order to achieve reliable quantitative results.

It is beyond the scope of this article to describe the various 3D reconstruction and segmentation procedures for different tomography methods and different materials. Instead, for illustration, we discuss the basic principles of 'qualitative image processing' (IP I) for a selected example (see Fig. 4.8). In this example, we consider a dataset from Pecho et al. [82, 83], which was acquired with FIB-SEM tomography from a fine-grained Ni-YSZ anode for solid oxide fuel cells (SOFC). Figure 4.8 a shows three orthogonal cross-sections of the original 3D raw data cube, before and after alignment. The raw data cube consists of 678 gray scale images with 2048×1768 pixels (i.e., $2.45 * 10^9$ voxels in total). The initial voxel resolution was $19.5 \times 19.5 \times 20$ nm (i.e., 19.5 nm pixel resolution in SEM-images (x–y plane) and 20 nm step size in FIB-sectioning (z-direction)). The size of the initial 3D image window (i.e., raw data volume) is thus $40 \times 34.5 \times 13.6$ μm ($1.87 * 10^4$ μm^3 in total).

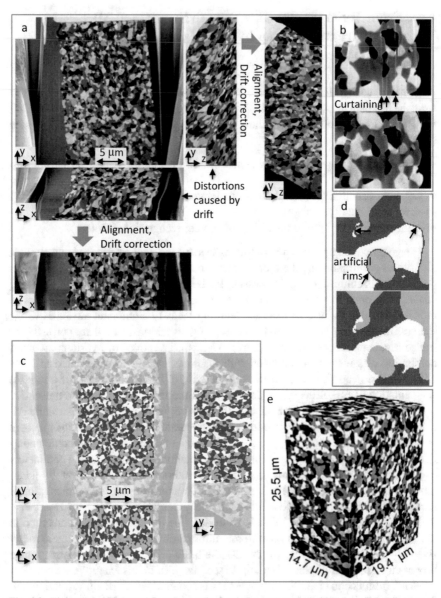

Fig. 4.8 Illustration of the workflow for qualitative image processing (IP I) for a FIB-SEM image stack. The processing includes filtering, 3D reconstruction and segmentation. The images represent the microstructure of a porous Ni-YSZ anode for SOFC [82, 83]. **a** 3D reconstruction of FIB-SEM raw data (stack of 2D images), before and after correction of drift in x, y- and z-directions, **b** correction of curtaining, **c** cropping region of interest (RoI) and segmentation into 3 phases: pores = white, nickel = green, YSZ = red, **d** removal of artificial rims (thin red line) at pore-nickel interface after threshold segmentation, **e** visualization of the final 3D microstructure model: pores = black, nickel = white, YSZ = gray

The raw data contains the following imperfections that need to be corrected before segmentation:

- noise caused by relatively fast acquisition rates,
- gray scale gradients typical for FIB-SEM images that are acquired under an angle of 52°,
- vertical stripes in y-direction (so called curtaining) caused by materials inhomogeneity and associated variation of the local ion milling rates,
- distortions in the image stack due to drift that could not be fully compensated during serial sectioning,
- brightness-flickering from image to image due to detector instabilities and/or charging.

For filtering, 3D-reconstruction and segmentation standard procedures were used, which are implemented in the commercial software GeoDict. Similar options are also available in other SW packages (see Table 4.3: e.g., Avizo, ImageJ, Fiji). In a first step, curtaining and flickering filters are applied for each 2D image of the stack (Fig. 4.8b). Then the 2D images are realigned so that the distortions caused by drift in x, y and z directions are corrected (as shown in Fig. 4.8a). After that, the reconstructed 3D volume is then resampled in order to obtain cubic voxels with edge lengths of 20 nm. Additional 3D image filters to reduce the noise and to increase the contrast are applied in a careful and conservative manner using a so-called non-local means (NLM) algorithm [161]. A suitable region of interest is then cropped (see the colored regions in Fig. 4.8c with 3D image window size of $17.28 \times 20.48 \times 12.96 \ \mu m$ = total $0.46 * 10^4 \ \mu m^3$, consisting of $864 \times 1024 \times 648$ voxels = total $0.73 * 10^9$ voxels). Finally, the gray scale volume is then segmented into the three major phases (nickel, YSZ and pores) by means of suitable threshold values obtained from histogram-analysis. Note that there exists a gray-scale gradient at the solid-pore interface due to the limited spatial resolution, which results from the non-finite beam-sample interaction volume (so-called excitation volume). Upon threshold segmentation, the intermediate gray levels at the interface between pores (black/white in c, d) and nickel (white / green in c, d) lead to artificial rims (gray / red in c, d) that are erroneously attributed to the YSZ phase (see thin red line in Fig. 4.8d top). Such erroneous rims represent a typical segmentation-artifact in three-phase materials, also described in [156]. They can be removed with a morphological opening operation, which consists of an erosion step followed by a dilation step (Fig. 4.8d bottom). The final 3D microstructure after filtering and segmentation is visualized in Fig. 4.8e). It represents a suitable input for quantitative image analysis (IP II) and numerical modeling.

4.5 Calculation Approaches for Tortuosity

This section describes methods for the computation of different tortuosity types. The underlying theories and concepts as well as the classification scheme and nomenclature were discussed in Chap. 2. More specific reviews on tortuosity calculation approaches were recently given by Tjaden et al. [162] and Fu et al. [163].

4.5.1 Calculation Approaches and SW for Direct Geometric Tortuosities (τ_{dir_geom})

In principle, almost all geometric tortuosities are based on the analysis of shortest pathways across the 3D microstructure in a predefined direction from inlet- to outlet-planes (i.e., $\tau_{dir\_geom} = L_{eff}/L_0$). Since, typically, there exist numerous shortest pathways connecting numerous couples of inlet- and outlet-points, the analysis of geometric tortuosity generally results in a histogram of paths lengths, from which a mean (effective) length (L_{eff}) with the corresponding mean tortuosity value can be determined. The crux is that the length of shortest pathways can be defined and measured in many different ways. Consequently, there exist various geometric tortuosities. The underlying principles and definitions for direct geometric tortuosities have already been discussed in Sect. 2.4. Here we only present a short summary of the corresponding calculation approaches and refer to suitable SW packages.

4.5.1.1 Direct Geodesic Tortuosity

The calculation approach for geodesic tortuosity ($\tau_{dir\_geodesic}$) is very simple and fast. The shortest pathways are defined in terms of the geodesic distance within the voxel space that represents the transporting phase [164]. Sometimes, this approach is *also called the direct shortest path searching method* (DSPSM) [163]. In the past, most authors dealing with geodesic tortuosity worked with in-house SW. Recently, an option for the computation of geodesic tortuosity was implemented in the commercial GeoDict software. The particular type of geodesic tortuosity introduced in [151] is implemented in the software package MIST [150]. Geodesic tortuosity currently takes a special role among the different geometric tortuosity types because it is used as a basis for empirical relationships between microstructure characteristics (porosity, tortuosity, constrictivity, hydraulic radius) and effective transport properties (see e.g., Stenzel et al. [164], Neumann et al. [165], and the discussion of empirical micro–macro relationships in Chap. 5). It was found in [164] that the geodesic tortuosity has a higher prediction power for estimating effective transport properties compared to medial axis tortuosity ($\tau_{dir\_medial\_axis}$).

4.5.1.2   Direct Medial Axis Tortuosity

The computation of the medial axis tortuosity ($\tau_{dir\_medial\_axis}$) is more complicated. It first requires the extraction of a medial axis skeleton [166]. Tortuosity is then computed from a set of shortest pathways along the medial axis skeleton, which are connecting couples of inlet- and outlet-points [64, 65]. Note that there exist many different skeletonization algorithms, which are for example implemented in dedicated modules of the software packages from Avizo/Amira (XPore Network) and/or Fiji (Skeletonize3D). Thereby, the resulting skeletons do not necessarily represent the medial axes. In this case, skeleton tortuosity ($\tau_{dir\_skeleton}$) is used as a more general term. Hence, to some degree, the resulting tortuosity values depend on the algorithm used for skeletonization, which leads to an additional complexity and uncertainty.

The different geometries of pathways for medial axis/skeleton tortuosity and geodesic tortuosity are illustrated and compared in Chap. 2 (see Figs. 2.2, 2.3 and 2.7). Empirical data from literature shows that $\tau_{dir\_medial\_axis}$ and $\tau_{dir\_skeleton}$ are usually not too different from each other, but they are consistently higher than $\tau_{dir\_geodesic}$ and τ_{dir_FMM}. This consistent order among the different tortuosity types was documented in Chap. 3 (see Figs. 3.6, 3.7 and 3.9).

4.5.1.3   Direct FMM Tortuosity

The fast-marching method tortuosity (τ_{dir_FMM}) is based on the simulation of a propagating front, which reveals the shortest geodesic pathways within the transporting phase [167–169]. This relatively simple calculation approach is thus very similar to the one used for the computation of geodesic tortuosity. Usually, in-house SW is used in order to determine τ_{dir_FMM}.

4.5.1.4   Direct PTM Tortuosity

The path tracking method tortuosity (τ_{dir_PTM}) can be considered as a fast and simple skeletonization approach, which however is only applicable for structures consisting of packed spheres. The algorithm identifies tetragons consisting of neighboring spheres. The pathways through the interstitial pores are found by connecting the gravity centers of adjacent tetragons in a predefined transport direction [170–172].

4.5.1.5   Direct Percolation Path Tortuosity

The percolation path tortuosity ($\tau_{dir\_percolation}$) is determined with an algorithm that allows the largest possible sphere(s) to travel along the shortest possible path from inlet- to outlet-plane. It should be noted that for a hypothetical case with very small 'spheres' (i.e., 1 pixel), the results for $\tau_{dir\_percolation}$ are identical to those for $\tau_{dir\_geodesic}$. However, when the largest possible sphere is considered, the narrow bottlenecks in the pore network hinder the direct passage of the sphere, which leads to longer pathways and higher tortuosity values compared to geodesic tortuosity. These different pathway-geometries are illustrated and compared in Fig. 2.7 (Chap. 2). The percolation path method is implemented for example in GeoDict, which allows the user to vary the range of sphere radii as well as the number of 'largest spheres pathways' to be analyzed (as optional input parameters). The larger the number of pathways, the smaller is the limiting sphere radius, and consequently, the smaller will be the corresponding tortuosity value. Hence, the method can be criticized since the results depend on the chosen parameters. Nevertheless, the percolation path method can be used for identification and visualization of transport pathways with a given, transport-limiting bottleneck size.

4.5.1.6   Direct Pore Centroid Tortuosity

The pore-centroid tortuosity ($\tau_{dir\_pore\_centroid}$) is a quantity, which can be computed by a quick and simple method that is based on determining the center of mass of the transporting phase (e.g., pores) in single 2D slices. The tortuous pathway is then tracked by connecting the mass centers of adjacent 2D slices in transport direction. This method is for example implemented in Avizo. It turns out that the obtained values decrease towards 1 when the volume fraction of the transporting phase increases, but also when the image window size increases (i.e., the center of gravity tends to be identical with the image center). Therefore, the relevance of the pore-centroid tortuosity is questionable.

4.5.2   Calculation Approaches and SW for Indirect Physics-Based Tortuosities (τ_{indir_phys})

Indirect, physics-based tortuosities (sometimes also called 'flux-based') are determined from effective transport properties, which are measured through specific transport experiments. These transport experiments can be performed either as a real physical experiment in the laboratory or as a virtual experiment by numerical simulation. A detailed discussion of literature dealing with laboratory experiments for electrochemical cells and for diffusion cells can be found in Tjaden et al. [162]. In the present section, however, we focus on the simulation-based approaches, which use 3D microstructure models from tomography as geometric input.

It is often mentioned that the mathematical treatment for different transport processes in numerical simulations is very similar and that the corresponding physics-based tortuosities can therefore be used interchangeably (i.e., it is assumed that $\tau_{indir\_ele} = \tau_{indir\_diff} = \tau_{indir\_therm}$). This hypothesis needs to be reevaluated critically. The aim is to understand which indirect tortuosities can or cannot be used interchangeably.

4.5.2.1 Comparison of Indirect Electrical, Diffusive, and Thermal Tortuosities

Materials characteristics and physical laws that are relevant for the simulation of different transport experiments are summarized in Table 4.4. We first consider the case of an electrical conduction experiment and its simulation, respectively. The liquid electrolyte in the pores acts as the transporting phase. The relevant intrinsic property is the electrical conductivity of the electrolyte (σ_0). In the simulation experiment a voltage difference between inlet and outlet planes is applied as driving force ($\Delta U/L$). The Laplace equation is solved under the assumption of charge conservation. At steady state conditions, the simulation reveals a constant electrical flux (J_{ele}). The effective conductivity (σ_{eff}) can then be calculated by substituting the simulated flux (J_{ele}) and associated voltage drop divided by the length of the simulation domain ($\Delta U/L$) in Ohm's law (Eq. 2.21, see Chap. 2). The resulting effective conductivity (σ_{eff}) is always smaller than the intrinsic conductivity (σ_0) due to the retarding effects from the materials microstructure. These retarding effects are generally attributed to the reduced pore volume fraction ($\varepsilon < 1$) and to the indirect electrical tortuosity (τ_{indir_ele}). Then, Eq. 2.24b ($\sigma_{eff} = \sigma_0\, \varepsilon/\tau_{ele}^2$) is usually taken as a quantitative description of the involved micro–macro relationship. Hence, knowing the porosity (ε) from image analysis and the effective conductivity (σ_{eff}) from simulation, the electrical tortuosity can be computed indirectly according to Eq. 2.25 ($\tau_{indir\_ele} = \sqrt{(\sigma_0\, \varepsilon/\sigma_{eff})}$).

As shown in Table 4.4, the material laws and the physical laws for thermal conduction and for bulk diffusion are very similar to those for the electrical conduction. Therefore, the simulation of these transport processes can be performed in a very similar way. From a mathematical point of view, Ohm's law, Fick's law, and Fourier's law reveal exactly the same relationship between the steady state fluxes (electric, diffusive or thermal fluxes), the effective properties (electric conductivity, diffusivity, thermal conductivity) and the applied driving forces (gradients of electric potential, of concentration, and of temperature).

Several authors [8, 81, 162, 163, 173, 174] performed comparative modeling studies using identical 3D microstructures as input for simulations of different transport processes (i.e., bulk diffusion as well as electrical and thermal conduction). These studies document that the three simulation approaches reveal exactly the same results for the relative properties (i.e., $X_{rel} = \sigma_{eff\_ele}/\sigma_{0\_ele}$ or D_{eff}/D_0 or $K_{eff\_thermal}/K_{0\_thermal}$) and consequently also for the corresponding indirect tortuosities ($\tau_{indir\_ele} = \tau_{indir\_diff} = \tau_{indir\_thermal}$). In these studies, the consistency of results could be demonstrated even for cases where different numerical methods were used (i.e., FVM, FDM, FVM, LBM and random walk). These findings indicate that, in principle, different numerical simulation approaches for diffusion and conduction are highly reproducible and thus, the corresponding indirect tortuosities for electric and thermal conduction as well as Fick's diffusion can be used interchangeably.

As an exception, it must be emphasized that diffusion in nano-porous media requires a different treatment of the microstructure effects. The so-called Knudsen

Table 4.4 Calculation of effective transport properties: summary of material properties and physical laws relevant for transport experiments and associated transport simulation approaches

Experiment	Electric (or ionic) conduction	Bulk diffusion	Thermal conduction	Viscous flow
Type of τ	$\tau_{indir\_ele\_sim}$	$\tau_{indir\_diff\_sim}$	$\tau_{indir\_therm\_sim}$	$\tau_{indir\_hydr\_sim}$
Transported species	Electrons (or ions)	Chemical species	Heat	Viscous medium
Transporting phase	Electrolyte in pore or a conductive solid	Gas or liquid in pore or solid	Gas or liquid in pore or solid	Gas or liquid in pore
Intrinsic property	Intr. electric (or ionic) conductivity (σ_0)	Intr. diffusivity (D_0)	Intr. thermal conductivity (K_0)	Viscosity (μ)
Effective property	Eff. electric (or ionic) conductivity (σ_{eff})	Eff. diffusivity (D_{eff})	Eff. thermal conductivity (K_{eff})	Permeability (κ)
Driving force	Voltage gradient ($\Delta U/L$)	Conc. gradient ($\Delta C/L$)	Temp. gradient ($\Delta T/L$)	Pressure grad. ($\Delta P/L$)
Transport law	Ohm's law $J_{ele} =$ $\sigma_{eff}\,(\Delta U/L)$ Eq. 2.21	Fick's law $J_{diff} =$ $D_{eff}\,(\Delta C/L)$ Eq. 2.29	Fourier's law $J_{therm} =$ $K_{eff}\,(\Delta T/L)$	Darcy's law (simple case) $J_{hydr} =$ $(\kappa/\mu\,(\Delta P/L)$ Navier–Stokes (general)
Micro–macro relation	$\sigma_{eff} = \sigma_0\,\varepsilon/\tau_{ele}^2$ Eq. 2.24b	$D_{eff} = D_0\,\varepsilon/\tau_{diff}^2$ Eq. 2.31	$K_{eff} =$ $K_0\,\varepsilon/\tau_{therm}^2$	$\kappa =$ $r_{hydr}^2\,\varepsilon/2\,\tau_{hydr}^2$ Eq. 2.9
Resolving τ	$\tau_{ele} =$ $\sqrt{(\sigma_0\,\varepsilon/\sigma_{eff})}$ Eq. 2.25	$\tau_{diff} =$ $\sqrt{(D_0\,\varepsilon/D_{eff})}$ Eq. 2.32	$\tau_{therm} =$ $\sqrt{(K_0\,\varepsilon/K_{eff})}$	$\tau_{hydr} =$ $\sqrt{(r_{hydr}^2\,\varepsilon/\kappa)}$ Eq. 2.16

diffusion, which was discussed earlier (see Eqs. 2.34–2.36), is then often simulated with a random walk approach.

In the context of indirect tortuosity, a critical point and a source of uncertainty is the underlying assumption of a known quantitative micro–macro relationship. It is often postulated that the micro–macro relationship in porous media can be described with simple expressions such as Eq. 2.24b for electrical conduction ($\sigma_{eff} = \sigma_0\,\varepsilon/\tau_{indir\_ele}^2$) and with analogous relationships for diffusivity and thermal conduction (e.g., Eq. 2.31: $D_{eff} = D_0\,\varepsilon/\sigma_{indir\_ele}^2$). This is, however, a very simplified assumption, whereby all resistive effects induced by the morphology of the microstructure are lumped together in the indirect tortuosity, except for the volume effect that is accounted for by ε (see Eq. 2.25: $\tau_{indir\_phys} = \sqrt{(\varepsilon/\sigma_{rel})}$). For the same 3D microstructures, this calculation approach typically results in indirect tortuosities that are much higher than the direct geometric and the mixed tortuosities (see Chap. 3, Fig. 3.9, relative order of tortuosity types).

It must be noted that various authors came up with alternative descriptions for the underlying micro–macro relationships. Some authors postulate a more *exclusive approach*, whereby the *bottleneck effect is removed from indirect tortuosity* [62, 63, 175–177]. This exclusion is achieved by adding a distinct constrictivity parameter (β) into the micro–macro relationships (i.e., $\tau_{indir\_phys} = \sqrt{(\varepsilon\,\beta/\sigma_{rel})}$, see also Eqs. 2.26, 2.27 and 2.33). As discussed in [63], the exclusive approach with separate treatment of constrictivity results in significantly lower values for the indirect tortuosity compared to the standard definition via Eqs. 2.24b and 2.25. The values obtained in this way for indirect tortuosity are then more similar to the values obtained for direct geometric tortuosity.

In contrast, in a more *inclusive approach*, some authors *added the pore volume effect to the indirect tortuosity* by removing ε from the equation (i.e., $\tau_{indir\_phys} = \sqrt{(1/\sigma_{rel})}$). Then, the corresponding property is sometimes called diffusibility instead of indirect tortuosity [178]. This inclusive approach leads to even higher values for the indirect tortuosity compared to the standard definition.

In summary, this discussion illustrates that the *indirect tortuosity heavily depends on the specification of the underlying micro–macro relationships*. Regardless which definition one choses, in any case the indirect tortuosity does not capture the true geometric paths lengths. The indirect tortuosity describes some kind of a microstructure resistance, which always requires a clear definition of the underlying micro–macro relationship. Most frequently, the indirect tortuosity is calculated with Eq. 2.25.

4.5.2.2   Indirect Hydraulic Tortuosity

A 3D numerical framework can be established for pore-scale simulation of viscous flow in a similar way as discussed above for electrical conductivity. In this framework, a pressure gradient ($\Delta P/L$) is externally applied as driving force (instead of a potential gradient). The transported species and the transporting phase are the viscous medium in the pores. The hydraulic flux (J_{hydr}) can be computed at steady-state conditions by solving the (Navier-) Stokes equation using different numerical approaches (e.g., FVM, FEM, LBM). Permeability (κ) can then be calculated by substituting the simulated hydraulic flux and the corresponding pressure gradient into Darcy's equation (see Table 4.4, see also Eq. 2.2).

Despite the obvious analogies with conduction and diffusion (Ohm's law, Fourier's law, Fick's law), there also exist some fundamental differences in the physical and mathematical description of viscous flow (Navier Stokes equations). An important difference concerns the nature of the effective properties (i.e., permeability versus conductivity and diffusivity). Permeability itself is a pure microstructure property. In contrast to conduction and diffusion, there is no analogy for 'intrinsic permeability'. The intrinsic flow property can be ascribed to viscosity. In principle, permeability (κ) is comparable with the relative properties of conduction and diffusion (i.e., $\kappa \approx (\sigma_{eff\_ele}/\sigma_{0\_ele}) \approx (D_{eff}/D_0)$). These relative properties are entirely dependent on the microstructure. Thereby, small values for the relative properties represent high transport resistances. Nevertheless, whereas relative conductivity and relative diffusivity are dimensionless properties, permeability has units of m^2. This indicates that the microstructure imposes different limitations to viscous flow compared to conduction and diffusion (which is also obvious from the different forms of differential equations that are used to describe these transport processes). For conduction and diffusion, the microstructure limitations associated with volume fraction, paths lengths and bottlenecks are described by the dimensionless characteristics of porosity (ε), tortuosity (τ) and constrictivity (β) (see e.g., Eqs. 2.24b, 2.26, 2.27, 2.31 and 2.33). For flow and permeability, there exists an additional microstructure effect, which is caused by viscous drag at the pore walls. As discussed in Chap. 2, this flow specific effect at the pore walls can be expressed with the squared hydraulic radius ($r_h{}^2$). Permeability is thus described by a combination of dimensionless characteristics (ε, τ) and a length-dependent characteristic (r_h) according to Eq. 2.9 ($\kappa\ r_h{}^2\ \varepsilon/\tau_{hydr}{}^2$). In principle, the hydraulic tortuosity can now be computed indirectly from Eq. 2.9 ($\tau_{indir\_hydr} = \sqrt{(r_h{}^2\varepsilon/\kappa)}$). However, to do so it is necessary to also assess the hydraulic radius (r_h), in addition to permeability and porosity. Until recently, suitable 3D image analysis methods for the measurement of hydraulic radius were lacking for complex microstructures. As discussed in Chap. 2, the Carman-Kozeny equations provide solutions that are valid only for simplified microstructures (packed spheres, parallel tubes). Novel methods of 3D image analysis to determine the hydraulic radius, which can be reliably computed for complex, disordered microstructures, will be discussed in Chap. 5. The lack of suitable methods for characterization of the hydraulic radius may be the main reason why indirect hydraulic tortuosity (τ_{indir_hydr}) has not been considered in previous studies of pore-scale flow. As an alternative approach, it is possible to measure hydraulic tortuosity from simulated 3D velocity fields (and associated streamlines). These types of hydraulic tortuosity (i.e., $\tau_{mixed\_hydr\_streamline}$, $\tau_{mixed\_hydr\_Vav}$) however belong to the class of mixed tortuosities and they contain completely different information than the indirect tortuosities (see Sect. 4.5.3).

For indirect tortuosities, it can be summarized that the *microstructure resistance is different for viscous flow compared to conduction and diffusion*. The computation of indirect hydraulic tortuosity related to flow is more complex and therefore it is hardly used. *The physics-based indirect tortuosities for electrical and thermal conduction and for bulk diffusion can be used interchangeably* (but not for Knudsen diffusion). In several comparative studies [8, 81, 147, 162, 173, 174] it was shown that different simulation approaches (FVM, FDM, random walk) and different voxel-based SW packages (TauFactor, Avizo, GeoDict, PyTrax) provide almost identical results for the indirect tortuosities of conduction and diffusion (τ_{indir_ele}, τ_{indir_diff}, $\tau_{indir\_thermal}$). Tjaden et al. [162] concluded that uncertainties and errors from segmentation and meshing are much more important than those from different simulation approaches.

Commercial and open-source SW packages for 3D numerical simulation of different kinds of transport (conduction, diffusion, flow) and for computation of associated indirect tortuosities are summarized in Table 4.3. For the characterization of complex microstructures, it is recommended to use SW packages that enable transport simulations with a precise (i.e., voxel-based) geometric representation of the microstructure, because this approach is usually more reliable than mesh-based approaches with a reduced number of elements. The voxel-based option is available for example in the SW packages GeoDict, PuMA, Avizo, Amira, PerGeos, Pore3D, TauFactor, Pytrax, OpenLB and Palabos.

4.5.3 Calculation Approaches for Mixed Tortuosities

Hydraulic tortuosity cannot easily be determined indirectly from effective properties as this is the case for electric or diffusive tortuosities (e.g., τ_{indir_ele}). An alternative approach to characterize hydraulic tortuosity focuses on streamlines representing the flow paths. This approach was discussed already in 1937 by Carman [179]. According to Eq. 2.17, the effective length of the hydraulic flow path is defined as weighted average of streamline lengths ($L_{eff\_weighted}$), from which the hydraulic tortuosity can then be deduced as follows

$$\tau_{mixed\_hydr\_streamline} = \frac{L_{eff\_weighted}}{L} = \frac{1}{L} \frac{\sum_i L_i w_i}{\sum_i w_i}. \qquad (4.5)$$

As discussed by various authors [180–185], the *definition and computation of suitable weighting factors (w_i) is a major challenge, which puts strong limitations to the practical use of streamline tortuosities.* As an alternative approach, it was shown by Matyka and Koza [152] and Duda et al. [186] that mixed hydraulic tortuosity can be computed in a much simpler way, based on the integration of local vector components from a simulated 3D velocity field. This so-called volume-averaged tortuosity was described in Chap. 2, Eq. 2.13 ($\tau_{mixed\_hydr\_Vav} = <v_c>/<v_x>$) and Eq. 2.18.

Hence, both, volume averaged as well as streamline tortuosities, require a *3D vector field from numerical flow simulation as a basis for the computation of mixed hydraulic tortuosity.* The underlying flow simulations can be performed with different numerical methods (FVM, FDM, FEM, LBM). Furthermore, these mixed tortuosities can be determined not only for viscous flow but also for other types of transport (i.e., conduction and diffusion), for which a 3D vector field can be computed. Hence, the *streamline and volume-averaged tortuosities are also physics- or flux-based tortuosities.* However, in contrast to the indirect physics-based tortuosities, streamline and volume-averaged tortuosities are *calculated by a geometric analysis of 3D vector fields* (and not from the effective property itself). The mixed tortuosities thus bear a higher level of information since they *combine physical and geometric information.* The mixed tortuosities are thus of major importance if one wants to understand the *true path length effects.* As discussed in Chap. 3 (compilation of empirical data), the values obtained for mixed tortuosities are in the same range as those from geometric tortuosities, but consistently lower than indirect tortuosities. This finding supports the interpretation that indirect tortuosities overestimate the limiting effects from tortuous pathways.

In literature, different modeling approaches are reported in order to obtain the required 3D vector fields for the calculation of mixed tortuosities. The numerical approach itself (i.e., differences between FVM, LBM etc.) has a lower impact on the resulting tortuosity than, e.g., inaccuracy from meshing or segmentation. In most studies, in-house solutions are used for the analysis of 3D vector fields and for determination of mixed tortuosities. For a detailed description of volume-averaged tortuosity and its implementation see Matyka and Koza [152]. Recently, the option for characterizing volume-averaged tortuosities ($\tau_{mixed\_hydr\_Vav}$, $\tau_{mixed\_diff\_Vav}$, $\tau_{mixed\_ele\_Vav}$, $\tau_{mixed\_therm\_Vav}$) by combining numerical transport simulations with 3D image analysis of the flow fields was implemented in the SW package GeoDict.

4.6   Pore Scale Modeling for Tortuosity Characterization: Examples from Literature

The *analysis of mixed and indirect tortuosities is based on pore scale modeling.* However, a review of transport modeling techniques and associated equations is beyond the scope of this publication. Instead, we refer to corresponding textbooks (see e.g., Bird et al. [187], Sahimi [188], Bear [189]). Furthermore, it must be emphasized that the phenomena of transport in porous media, the corresponding pore scale modeling and the associated analysis of tortuosity can be very complex. Depending on the system under consideration, complexity can be introduced for example by coupling of a standard transport process (e.g., diffusion or flow) with additional processes such as electrochemical reactions, physical interactions at pore walls (Knudsen effect, adsorption etc.) and/or reactive transport (chemical interaction with solids). Also, the simulation of transport phenomena at different length scales is often an important issue. It is also beyond the scope of this review to discuss the impact of such complexities on tortuosity. Instead, we present some examples from literature with different modeling approaches for material systems with complex transport phenomena.

4.6.1   Examples of Pore Scale Modeling in Geoscience

Saxena et al. [190] define a benchmark with numerous 3D-microstructure models, which are used for comparison of different flow simulation codes (LBM; mesh-based FEM and openFoam; voxel-based FFT and stokes-LIR).

Su et al. [191] describe various methods for pore-scale simulation (2 phase flow), including the pore network model, LBM, Navier–Stokes equation-based interface tracking methods, and smoothed particle hydrodynamics.

Liu et al. [29] present a critical review on computational challenges in petrophysics using micro-CT and up-scaling.

He et al. [192] perform molecular dynamics (MD) simulations of gas diffusion in nano-porous shale in order to evaluate diffusive tortuosity.

Wang et al. [193] present a review of analytical and semi-analytical fluid flow models for ultra-tight hydrocarbon reservoir rocks (including fracking).

Müter et al. [194] simulate diffusion in nano-scale pore networks based on dissipative particle dynamics (DPD).

Tallarek et al. [195] present a multi-scale simulation approach for diffusion in porous media. It covers interfacial dynamics at molecular scale as well as hierarchical porosity at meso- and macro-scales.

Ghanbarian [196] discussed the problem of scale dependency in rocks and soils, which results in scattered plots of tortuosity and diffusion coefficient versus scales. By applying finite-size scaling analysis the data show a quasi-universal trend.

4.6.2   Examples of Pore Scale Modeling for Energy and Electrochemistry Applications

Ryan and Muckerjee [197] give a critical overview on pore-scale modeling approaches for electrochemical devices (i.e., fuel cells and batteries). Particular focus is given to direct numerical simulation (DNS) techniques, which includes particle-based methods (smoothed particle hydrodynamics, dissipative particle dynamics, LBM) and fine-scale CFD methods (voxel-based vs. mesh-based).

Usseglio-Viretta et al. [198] demonstrated how to resolve the discrepancy in tortuosity factor estimation for Li-ion battery electrodes based on a combination of micro- and macro-modeling with experimental characterization.

Lu et al. [199] discuss the concept of digital microstructure design for lithium-ion battery electrodes based on a combination of nano-CT and multi-physics modeling.

Le Houx and Kramer [200] present a review on physics-based modeling of porous lithium-ion battery electrodes.

Zhang et al. [201] describe an experimentally validated pore-scale Lattice Boltzmann model to simulate the performance of redox flow batteries.

Recent publications dealing with the modeling of porous electrodes, the complex transport phenomena at pore scale (Chen et al. [202]), as well as multi-scale phenomena of ion transport (Tao et al. [203]) are addressed specifically.

Fundamental aspects of SOFC modeling such as coupled electrochemistry and transport at micro- to meso-scales as well as impedance analysis are reviewed by Grew and Chiu [204], Hanna et al. [205], Dierickx et al. [206] and Timurkutluk et al. [207].

Models for PEM fuel cells are discussed by Weber et al. [208] (review of transport models), Zenyuk et al. [209] (coupling of pore- and continuum-scales/up-scaling) and Liu et al. [210] (Liquid water transport in GDL).

4.7   Stochastic Microstructure Modeling

Statistical analysis of microstructure effects, for example in studies that aim for establishing quantitative relationships between tortuosity, porosity, and effective transport properties, is generally limited by the availability of suitable 3D image data. In a conventional approach using experimental materials fabrication followed by tomography and image analysis, the number of 3D analyses that can be performed with reasonable effort (in time and money) is usually quite limited. In this context, stochastic microstructure modeling is a powerful method that offers the possibility to increase the amount of 3D image data efficiently. Microstructure modeling is thus particularly important for *data driven, statistical investigations of microstructure effects.*

Stochastic geometry (also called mathematical morphology by some authors) represents the *mathematical basis for stochastic microstructure modeling.* Overviews related to the use of stochastic geometry for microstructure modeling are given by Chiu et al. [144], Matheron [211], Jeulin [212], Lantéjoul [213] and Schmidt [214]. In principle, stochastic geometry provides a mathematical toolbox for the generation of virtual, but realistic microstructures, which consists of many different approaches: random point processes, random closed sets, surface processes, random tessellations as well as random geometrical graphs representing spatial networks. The challenge of microstructure modeling is to use appropriate tools of stochastic geometry and mathematical morphology to develop stochastic 3D microstructure models, which allow for the generation of digital twins of a specific microstructure. In principle, two main *quality criteria* must be fulfilled:

Prediction power

A suitably chosen stochastic model provides virtual 3D microstructures, based on which the *structural properties (e.g., tortuosity) and performances* (e.g., transport resistance) of real materials can be *predicted with* a *high precision and reliability.*

Efficiency

The *generation of virtual 3D microstructures* with a stochastic model must be efficient in order to *enable extensive parameter sweeps* for data driven, statistical investigations.

An extensive literature review of microstructure modeling from a materials science perspective is given by Bargmann et al. [215]. In this review, different approaches for stochastic microstructure modelling are discussed in context with the type of microstructure for which these models are suitable (Table 4.5).

In [215], two main *approaches for microstructure modeling* are distinguished:

Physics-based methods

The physics-based methods aim for the simulation of physical processes liable for microstructure formation. For example, with the phase-field method, physical

Table 4.5 Classification scheme for different types of microstructures, modified after Bargmann et al. [215]

1. Porous materials	(Application examples)
1.1. Agglomerates (consolidated particles)	
1.1.1 Cellular structures	Lightweight materials
1.1.1.1 Open cells	Metallic and ceramic foams
1.1.1.2 Closed cells	Closed cell polyurethane (CCPU)
1.1.2 Granular materials with interstitial pore networks	
1.1.2.1 Two-phase materials	Sandstones, battery electrodes
1.1.2.2 Three-phase materials	Cermet anodes for SOFC
1.1.3 Dilute voids (porosity below percolation)	Low porosity rocks (clays), dense slag
1.2 Fabrics and fiber-based materials	
1.2.1 Woven fabrics	Textiles
1.2.2 Non-woven fabrics	Filter materials, PEM GDL
1.3 Aggregates (non-consolidated particles)	Sands, soils, powders, packed spheres
2. Non-porous materials	
2.1 Polycrystalline materials	Alloys and dense ceramics
2.1.1 Granular structures	Al-Sn alloy
2.1.2 Lamellar structures	Ti-aluminide with $\alpha + \gamma$-phases
2.2 Bi-continuous composites (BC)	
2.2.1 Random BC	Cermets: ceramic–metal composites
2.2.2 Regular BC	Block co-polymers
2.3 Matrix-inclusion composites (MIC)	Carbide particles in metal matrix
2.3.1 Particle reinforced MIC	Carbon black in rubber
2.3.2 Fiber reinforced MIC	Polymer matrix-fiber composites

processes of grain growth or crystallization and associated microstructure formation are described with so-called transformation rules.

Geometrical methods

The geometrical methods aim for mimicking the material's morphology disregarding the underlying physics of microstructure development. A prominent example for this approach is the random packing of particles (spheres, ellipsoids, polyhedron, cylinders, fibers etc.) by means of discrete element modeling (DEM), see e.g., Sheikh and Pak [216].

For almost any type of microstructure, descriptions of suitable models can be found in literature with both, geometrical as well as physics-based approaches. However, only a few SW packages are available for microstructure modeling by stochastic geometry, as shown in Table 4.3. ESyS, GenGeo, YADE and Mote3D represent a small group of dedicated SW packages for microstructure generation, which are based e.g., on discrete element modeling (DEM). Further SW packages with specific modules for microstructure generation are GeoDict (e.g., GrainGeo, FiberGeo), Digimat, and PuMa. The BruggemanEstimator provides a module for the generation of battery structures and Dream3D for simulation of crystalline grain orientation patterns in the context with 3D EBSD (i.e., FIB-SEM tomography).

In the following sections we present short reviews from two important application fields of stochastic microstructure modeling, which are digital materials design for electrochemical devices and digital rock physics (DRP). For stochastic microstructure modeling of cellular and foam materials, see e.g., [217–220].

4.7.1   Stochastic Modeling for Digital Materials Design (DMD) of Electrochemical Devices

An *overview of microstructure modeling approaches for electrochemical devices* can be found in Ryan and Mukherjee [197]. Thereby, different stochastic 3D reconstruction methods are presented, which include Monte Carlo modeling, dynamic particle packing, stochastic grids, simulated annealing and controlled random generation. These stochastic models enable the creation of various 3D microstructures that are important for batteries, PEM fuel cells and SOFCs.

Various stochastic models for electrochemical devices and energy materials have been presented in the literature, for example for the fibrous GDL in PEM fuel cells [221, 222], for granular microstructures of battery electrodes [99, 223–230], and for different types of SOFC electrodes [231–235]. Thereby, a particular challenge for the simulation of cermet anodes is the realistic description of connectivity in all three co-existing phases. This challenge can be solved with specific random geometric graphs,—so-called beta-skeletons [232].

In order to *discuss stochastic modeling as a basis for digital materials design (DMD)*, we consider an example of cermet anodes for SOFC, which is illustrated in Fig. 4.9. The aim of this approach is to optimize the anode microstructure, which consists of three phases, namely nickel, YSZ (zirconium oxide) and pore phases. Tomography data of this example is taken from [82, 83]. In an experimental approach, there are typically 3 to 5 main fabrication parameters, which can be used to vary the microstructure of cermet anodes. These parameters are related to composition (Ni/YSZ-ratio, pore former content), grain size of raw materials (powder fineness of Ni-oxide and YSZ) and sintering conditions (temperature, duration and pO_2 of gas environment). In order to find an optimized microstructure, it is necessary to perform systematic parametric sweeps, which results in a rather large test matrix. With a conventional experimental approach, this test matrix cannot be covered with a reasonable number of resources. Thus, in most studies, only a few samples can be investigated for example by means of FIB-SEM tomography and quantitative image analysis.

On the other hand, using a modern approach of digital materials design (DMD), the statistical basis is enlarged with the help of stochastic microstructure modeling.

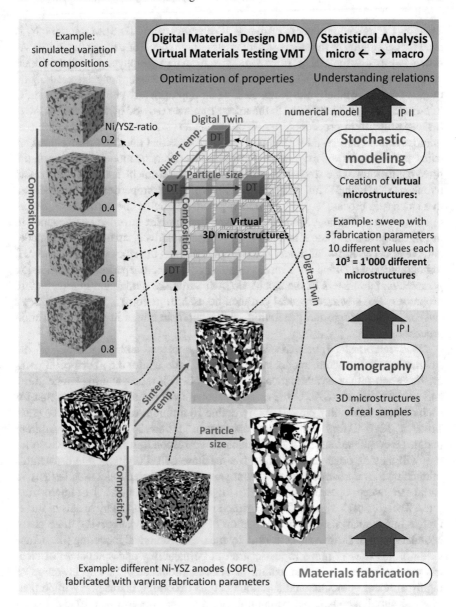

Fig. 4.9 Schematic illustration of the workflow for digital materials design (DMD) of SOFC anodes based on stochastic microstructure modeling. The stochastic model for the creation of numerous virtual anode microstructures is fitted to experimental tomography data (i.e., digital twins (DT) of real anode microstructures), which is why the virtual microstructures have realistic properties (3D-data taken from Pecho et al. [82, 83])

Thereby, the limited number of available 3D datasets from FIB-SEM tomography is used for calibration of the stochastic microstructure model (i.e., fitting to reality). As shown in Fig. 4.9, the real 3D microstructures from tomography represent the corner stones of an extended virtual parameter space. For each tomography dataset a digital twin is created, whereby the parameters of an appropriately chosen stochastic 3D microstructure model are fitted to the microstructure resolved by 3D imaging. After a successful fitting procedure, the digital twins are statistically similar to the real microstructures observed by 3D imaging. This means that microstructure character-istics and effective properties coincide nicely. Given that the model type is fixed, the fitted parameters of the stochastic microstructure model can be understood as 'rules' by means of which 3D microstructures with predefined properties can be produced in a stochastic process, e.g., by randomly placing particles of a certain size, shape, orientation in a 3D image volume. Moreover, the complex information contained in the 3D image data of microstructures is reduced to a relatively small number of model parameters. The fitting of model parameters calibrates the stochastic model to real tomography data. Doing so, a link is established between experimental fabrication parameters, parameters of the stochastic microstructure model and microstructure properties. The stochastic model can then be used to perform extensive parameter sweeps, which mimic the generation of 3D microstructures in a real fabrication process.

Once relationships between fabrication parameters and parameters of the stochastic microstructure model are established, one could, e.g., perform a para-metric sweep with three fabrication parameters (Ni-YSZ ratio, sintering temperature, and particle size of YSZ). In the example shown in Fig. 4.9, it is assumed that for each of these fabrication parameters we define 10 different values (e.g., 1100, 1120, 1140, …. 1280 °C for the sintering temperature). This sweep results in 10^3 different parameter combinations, for each of which the corresponding virtual 3D microstruc-ture will then be created. This leads to a database of 1000 virtual microstructures mimicking the microstructure of differently manufactured electrodes. Such extensive parameter sweeps open new possibilities for data driven optimization of microstruc-tures. Thereby, each virtual 3D microstructure must be analyzed by means of quan-titative image analysis (i.e., determination of tortuosity, constrictivity, three phase boundary length, surface area etc.) and by means of numerical modeling (i.e., simu-lation of gas flow in pores for estimation of permeability and simulation of elec-tric conduction in Ni for estimation of effective conductivity). The invention of highly efficient computational solutions is thus becoming increasingly important. Massive simultaneous cloud computing (MSCC) and the use of artificial intelli-gence are promising techniques to solve the future challenges of big data analysis for DMD. Thus, novel concepts combining various techniques for modeling and computing represent the methodological basis at the forefront of innovative digital materials research. Thereby, stochastic microstructure modeling is identified as a key technology for digital materials science.

4.7.2 Stochastic Modeling for Digital Rock Physics and Virtual Materials Testing of Porous Media

The progress of 3D imaging, analysis and modeling also opens new possibilities to establish quantitative relationships between morphological microstructure characteristics (e.g., tortuosity, constrictivity, porosity etc.) and effective properties (permeability, diffusivity, strength, elasticity etc.) by means of virtual materials testing (VMT). Thereby, stochastic microstructure modeling provides the statistical basis for such data-driven investigations of microstructure effects. In geoscience this approach is often called digital rocks physics (DRP [236]).

For example, Berg 2012 and 2014 [176, 177] applied DRP to investigate the impact of tortuosity and other microstructure effects on conductivity and permeability of porous rocks, where the investigations are based on 5 micro-CT scans from Bentheimer (1 scan) and Fontainebleau (4 scans) sandstones. A numerical rock model, called e-Core, was used to create additional virtual sandstone microstructures (12 for Bentheimer and 7 for Fontainebleau sandstone) with different porosities, but with realistic properties. In this way, the micro–macro relationships in those sandstones could be described for materials with varying porosities, which is based on a set of 24 microstructures from micro-CT scans and virtual 3D-models.

In the meanwhile, a large amount of 3D image data including CT scans from real samples as well as virtual microstructures from stochastic modeling is available for free download from the 'Digital Rocks Portal' [237]. The latter is a public repository focusing on 3D microstructure data of porous media in geoscience. Such data can be used to perform data-driven investigations of micro–macro relationships with a broader data-basis.

Saxena et al. [190] generated a reference dataset consisting of a large variety of 3D microstructures ranging from idealized pipes to realistic digital rocks. The 3D models are used as a benchmark for DRP and for the comparison of different numerical solvers that are used in pore scale simulations (LBM, CFD, voxel based FDM, mesh-based FEM).

Nowadays, also commercial software for digital rocks physics has become available, such as PerGeos or GeoDict (see Table 4.3). These SW packages provide integrated solutions for the entire DRP-workflow, including 3D reconstruction, image analysis and numerical simulation. All these modern tools (including 3D data from repositories, stochastic models, SW packages for image analysis and numerical pore scale modeling) are increasingly used in different combinations for statistical analysis of microstructure effects (see e.g., Fu et al. [163]). The availability of the above-mentioned SW packages and their application to 3D image data representing a vast range of different microstructures will significantly contribute to a better understanding of the different types of tortuosities and of their relationship with effective transport properties in porous media.

In this context, we also refer to the recent work of Prifling et al. [238], where relationships between descriptors of two-phase microstructures, consisting of solid and pores, and their mass transport properties, have been investigated. To that end, a vast database has been generated comprising 90,000 porous microstructures drawn from nine different stochastic models, and their effective diffusivity and permeability as well as various microstructural descriptors have been computed. To the best of our knowledge, this is the largest and most diverse dataset created so far for studying the influence of 3D microstructure on mass transport in porous materials. The microstructures, descriptors, and the code used to study microstructure-property relationships are available open access via the following Zenodo repository: https://zenodo.org/record/4047774.

4.8 Summary

3D microstructure data is the basis for characterization of all three tortuosity categories, i.e., direct geometric, indirect physics-based and mixed tortuosities. The workflows for these 3 categories are illustrated in Fig. 4.1. For each step in these workflows, the underlying principles as well as current trends of methodologies are reviewed.

3D imaging

Nowadays, there are numerous 3D imaging methods available. We consider the four most important techniques, which are:

- *X-ray computed tomography* (including micro-CT, nano-CT, transmission and scanning X-ray microscopy (TXM, SXM)),
- *Serial sectioning techniques* (including FIB-SEM, PFIB-SEM, BIB-SEM, pulsed laser, Ultra-Microtom and mechanical polishing),
- *Electron tomography* (including 3D TEM and 3D STEM) and,
- Atom probe tomography (APT).

All these imaging techniques are rapidly evolving and improving. Today, the resolutions of these methods cover the lengths-scales from macroscopic scale down to atomic resolution. Depending on the imaging method there are also numerous detection and contrast modes available, which provide microstructure, chemical, crystallographic information and more. Of particular importance is the impressive improvement of time resolution in X-ray imaging, which opens new possibilities for 4D tomography at sub-μm resolution in combination with in-situ and in-operando experiments. In summary, when speaking about 3D imaging and tomography, it must be realized that we are dealing with a very versatile group of methodologies, which continues to make fast progress in various directions including improvement of resolution, acquisition time, detection mode, user friendliness etc. For materials scientist, the question is thus no longer, 'is there a suitable method available for characterization of my material?', but rather 'which method is suitable for characterization of my material?'

SW packages available for image processing and determination of tortuosity

After acquisition of 3D images, the raw data must be processed qualitatively and quantitatively. An extensive list with 75 SW packages is presented in Table 4.3.

The SW packages are grouped according to their modules and to the *options that they offer for*

- qualitative image processing (IP I: 3D-reconstruction, filtering, segmentation),
- quantitative image processing (IP II: various tortuosity types and other morphological characteristics),
- stochastic microstructure modeling and,
- numerical simulations (voxel-based vs mesh-based, transport vs multi-physics).

An example is presented, which illustrates the workflow for tortuosity characterization in SOFC electrodes based on FIB-SEM tomography and using the GeoDict SW.

Calculation approaches for tortuosity

For the *direct geometric tortuosities*, various types (i.e., geodesic, medial axis, FMM, PTM, percolation path, pore centroid) can be determined directly from the segmented 3D image data. For most types, suitable SW codes are available.

The *indirect physics-based tortuosities* are calculated from effective transport properties (i.e., electric or thermal conductivity, diffusivity or permeability). These effective properties can be determined either by transport simulation or with dedicated experiments. The comparison of involved mathematical and physical laws shows that electric and thermal conduction as well as bulk bulk diffusion are basically identical with each other, and therefore the corresponding indirect electrical, thermal and diffusional tortuosities can be used interchangeably. In contrast, the underlying physical and mathematical laws for viscous flow and associated permeability are different. The indirect hydraulic tortuosity can therefore not be used interchangeably with the other indirect tortuosities related to conduction and diffusion. Due to the indirect calculation approach, this tortuosity category must be *interpreted as a bulk measure for the transport resistance*, which includes various microstructure limitations (not only the path length effect). This explains, why the estimated values for indirect tortuosities are consistently higher than those for direct geometric and mixed tortuosities.

The procedure to calculate *mixed tortuosities* (e.g., streamline and volume averaged tortuosities) includes two main steps: *simulation of transport* (i.e., conduction, diffusion, or flow) and *determination of mean path length* based on geometric analysis of the simulated flow fields. Hence, mixed tortuosities include both, *physics-based information*, as well as *geometric information on the path lengths*. It must be emphasized that the mixed volume averaged tortuosity can be computed simply by integration of the local vector components, which is an elegant, efficient and reliable method. The *volume averaged tortuosity* is thus considered as the *most accurate approach to determine the true path lengths effect*.

Pore scale modeling

Transport simulations are the basis for calculation of indirect and mixed tortuosities. For a review of modeling techniques and mathematical equations describing transport in porous media we refer to existing textbooks (see e.g., Bird et al. [187], Sahimi [188], Bear [189]). For many applications with porous media, the transport phenomena can become rather complicated, e.g. due to coupled processes (e.g. reactive transport or poromechanics). Examples from literature dealing with such complex transport phenomena are presented for applications in geoscience and in electrochemistry. However, for standard cases the transport can be simply simulated with one of the above mentioned SW packages (Table 4.3).

Stochastic microstructure modeling

Microstructure modeling by means of stochastic geometry or discrete element modeling (DEM) is a powerful method that offers the possibility to increase the amount of 3D image data efficiently. Microstructure modeling is thus particularly important for *data driven, statistical investigations of microstructure effects.*

Two main approaches must be distinguished. With the *physics-based approach*, microstructures are created based on the simulation of the involved physical processes (e.g., crystallization, grain growth or mechanical deformation). With the *geometric approach* the microstructure is created so that it matches the morphological properties of a real microstructure independent from the physical process and associated history of the material.

Numerous strategies and codes for microstructure modeling are described in literature. The crux is to find a suitable method, which allows for an efficient microstructure realization that matches with the real microstructure properties of the investigated material. Examples are presented, where microstructure modeling is applied in the framework of digital materials design (DMD) for materials in electrochemical devices, as well as in digital rock physics (DRP).

References

1. J. Dirrenberger, S. Forest, D. Jeulin, Towards gigantic RVE sizes for 3D stochastic fibrous networks. Int. J. Solids Struct. **51**, 359 (2014)
2. T. Kanit, S. Forest, I. Galliet, V. Mounoury, D. Jeulin, Determination of the Size of the Representative Volume Element for Random Composites: Statistical and Numerical Approach. Int. J. Solids Struct. **40**, 3647 (2003)
3. M.D. Uchic, L. Holzer, B.J. Inkson, E.L. Principe, P. Munroe, Three-dimensional microstructural characterization using focused ion beam tomography. MRS Bull. **32**, 408 (2007)
4. C. Cao, M.F. Toney, T.-K. Sham, R. Harder, P.R. Shearing, X. Xiao, J. Wang, Emerging X-ray imaging technologies for energy materials. Mater. Today **34**, 132 (2020)
5. T.M.M. Heenan, D.P. Finegan, B. Tjaden, X. Lu, F. Iacoviello, J. Millichamp, D.J.L. Brett, P.R. Shearing, 4D nano-tomography of electrochemical energy devices using lab-based X-ray imaging. Nano Energy **47**, 556 (2018)
6. T.M.M. Heenan, C. Tan, J. Hack, D.J.L. Brett, P.R. Shearing, developments in X-ray tomography characterization for electrochemical devices. Mater. Today (2019)

7. T.M.M. Heenan, C.Tan, A.J. Wade, R. Jervis, D.J.L. Brett, P.R. Shearing, Theoretical transmissions for X-ray computed tomography studies of lithium-ion battery cathodes. Mater. Des. **191** (2020)

8. F. Tariq, V. Yufit, M. Kishimoto, P.R. Shearing, S. Menkin, D. Golodnitsky, J. Gelb, E. Peled, N.P. Brandon, Three-dimensional high resolution X-ray imaging and quantification of lithium ion battery mesocarbon microbead anodes. J. Power Sources **248**, 1014 (2014)

9. A. Bertei, E. Ruiz-Trejo, F. Tariq, V. Yufit, A. Atkinson, N.P. Brandon, Validation of a physically-based solid oxide fuel cell anode model combining 3D tomography and impedance spectroscopy. Int. J. Hydrogen Energy **41**, 22381 (2016)

10. F. Tariq, P.R. Shearing, R.S. Bradley, J. Gelb, P.J. Withers, N.P. Brandon, *4D Tomography : Imaging of Microstructural Evolution in Fuel Cells Using High Resolution X-Ray Tomography*, vol 1 (n.d.)

11. F. Tariq, R. Haswell, P.D. Lee, D.W. McComb, Characterization of hierarchical pore structures in ceramics using multiscale tomography. Acta Mater. **59**, 2109 (2011)

12. O.O. Taiwo, M. Loveridge, S.D. Beattie, D.P. Finegan, R. Bhagat, D.J.L. Brett, P.R. Shearing, Investigation of cycling-induced microstructural degradation in silicon-based electrodes in lithium-ion batteries using X-ray nanotomography. Electrochim. Acta **253**, 85 (2017)

13. M. Wolf, B.M. May, J. Cabana, Visualization of electrochemical reactions in battery materials with X-ray microscopy and mapping. Chem. Mater. **29**, 3347 (2017)

14. S.L. Morelly, J. Gelb, F. Iacoviello, P.R. Shearing, S.J. Harris, N.J. Alvarez, M.H. Tang, Three-dimensional visualization of conductive domains in battery electrodes with contrast-enhancing nanoparticles. ACS Appl. Energy Mater **1**, 4479 (2018)

15. S.R. Daemi et al., 4D visualisation of in situ nano-compression of Li-ion cathode materials to mimic early stage calendering. Mater. Horiz. **6**, 612 (2019)

16. Q. Meyer, J. Hack, N. Mansor, F. Iacoviello, J.J. Bailey, P.R. Shearing, D.J.L. Brett, Multi-scale imaging of polymer electrolyte fuel cells using x-ray micro- and nano-computed tomography. Trans. Electron. Microsc. Helium-Ion Microsc. Fuel Cells **19**, 35 (2019)

17. N. Kulkarni, M.D.R. Kok, R. Jervis, F. Iacoviello, Q. Meyer, P.R. Shearing, D.J.L. Brett, The effect of non-uniform compression and flow-field arrangements on membrane electrode assemblies—X-ray computed tomography characterisation and effective parameter determination. J. Power Sources **426**, 97 (2019)

18. Y. Nagai, J. Eller, T. Hatanaka, S. Yamaguchi, S. Kato, A. Kato, F. Marone, H. Xu, F.N. Büchi, Improving water management in fuel cells through microporous layer modifications: fast operando tomographic imaging of liquid water. J. Power Sources **435**, 226809 (2019)

19. M. Bührer, M. Stampanoni, X. Rochet, F. Büchi, J. Eller, F. Marone, High-numerical-aperture macroscope optics for time-resolved experiments. J. Synchrotron Radiat **26**, 1161 (2019)

20. A. Mularczyk, Q. Lin, M. J. Blunt, A. Lamibrac, F. Marone, T.J. Schmidt, F.N. Buchi, J. Eller, Droplet and percolation network interactions in a fuel cell gas diffusion layer. J. Electrochem. Soc. (2020)

21. S. Brisard, M. Serdar, P.J.M. Monteiro, Multiscale X-ray tomography of cementitious materials: a review. Cem. Concr. Res. **128**, 105824 (2020)

22. A. du Plessis, W.P. Boshoff, A review of X-ray computed tomography of concrete and asphalt construction materials. Constr. Build. Mater. **199**, 637 (2019)

23. S.C. Garcea, Y. Wang, P.J. Withers, X-ray computed tomography of polymer composites. Compos. Sci. Technol. **156**, 305 (2018)

24. A. Du Plessis, I. Yadroitsev, I. Yadroitsava, S.G. Le Roux, X-ray microcomputed tomography in additive manufacturing: a review of the current technology and applications. 3D Print Addit. Manuf. **5**, 227 (2018)

25. F. Iacoviello, X. Lu, T.M. Mitchell, D.J.L. Brett, P.R. Shearing, The imaging resolution and knudsen effect on the mass transport of shale gas assisted by multi-length scale X-ray computed tomography. Sci. Rep. **9**, 1 (2019)

26. T. Bultreys, W. De Boever, V. Cnudde, Imaging and image-based fluid transport modeling at the pore scale in geological materials: a practical introduction to the current state-of-the-art. Earth Sci. Rev. **155**, 93 (2016)

27. T. Bultreys, M.A. Boone, M.N. Boone, T. De Schryver, B. Masschaele, L. Van Hoore-beke, V. Cnudde, Fast laboratory-based micro-computed tomography for pore-scale research: illustrative experiments and perspectives on the future. Adv. Water Resour. **95**, 341 (2016)

28. V. Cnudde, M.N. Boone, High-resolution X-ray computed tomography in geosciences: a review of the current technology and applications. Earth Sci. Rev. **123**, 1 (2013)

29. J. Liu, G.G. Pereira, Q. Liu, K. Regenauer-Lieb, Computational challenges in the analyses of petrophysics using microtomography and upscaling: a review. Comput. Geosci. **89**, 107 (2016)

30. S. Peng, F. Marone, and S. Dultz, Resolution effect in X-ray microcomputed tomography imaging and small pore's contribution to permeability for a berea sandstone. J. Hydrol. (Amst) **510**, 403 (2014)

31. P. Zhang, Y. Il Lee, J. Zhang, A review of high-resolution x-ray computed tomography applied to petroleum geology and a case study. Micron **124**, 102702 (2019)

32. R. Ditscherlein, O. Furat, E. Löwer, R. Mehnert, R. Trunk, T. Leißner, M.J. Krause, V. Schmidt, U.A. Peuker, PARROT: a pilot study on the open access provision of particle-discrete tomographic datasets. Microsc. Microanal. **1** (2022)

33. S.D. Rawson, J. Maksimcuka, P.J. Withers, S.H. Cartmell, X-ray computed tomography in life sciences. BMC Biol. **18**, 1 (2020)

34. H. Yan, P.W. Voorhees, H.L. Xin, Nanoscale X-ray and electron tomography. MRS Bull. **45**, 264 (2020)

35. A.P. Cocco, G.J. Nelson, W.M. Harris, A. Nakajo, T.D. Myles, A.M. Kiss, J.J. Lombardo, W.K.S. Chiu, Three-dimensional microstructural imaging methods for energy materials. Phys. Chem. Chem. Phys. **15**, 16377 (2013)

36. E. Maire, P.J. Withers, Quantitative X-ray tomography. Int. Mater. Rev. **59**, 1 (2014)

37. P. Pietsch, V. Wood, X-ray tomography for lithium ion battery research: a practical guide. Annu. Rev. Mater. Res. **47**, 451 (2017)

38. Zeiss, An overview of 3D X-ray microscopy, Essential Knowledge Briefings **1**, 1 (2020)

39. J. Gondzio, M. Lassas, S.-M. Latva-Äijö, S. Siltanen, F. Zanetti, Material-separating regularizer for multi-energy X-ray tomography. Inverse Probl. **38**, 025013 (2022)

40. E. Maire, C. Le Bourlot, J. Adrien, A. Mortensen, R. Mokso, 20 Hz X-ray tomography during an in situ tensile test. Int. J. Fract. **200**, 3 (2016)

41. R. Mokso et al., GigaFRoST: the gigabit fast readout system for tomography. J. Synchrotron Radiat **24**, 1250 (2017)

42. F. De Carlo et al., TomoBank: a tomographic data repository for computational x-ray science. Meas. Sci. Technol. **29** (2018)

43. K. Bugelnig, P. Barriobero-Vila, G. Requena, Synchrotron computer tomography as a characterization method for engineering materials. Prakt. Metallogr./Pract. Metallogr. **55**, 556 (2018)

44. D. Kazantsev, E. Guo, A. B. Phillion, P.J. Withers, P.D. Lee, Model-based iterative reconstruction using higher-order regularization of dynamic synchrotron data, Meas. Sci. Technol. **28** (2017)

45. M. Hidayetoglu, T. Bicer, S.G. de Gonzalo, B. Ren, D. Gursoy, R. Kettimuthu, I.T. Foster, W.-M.W. Hwu, MemXCT: design, optimization, scaling, and reproducibility of x-ray tomography imaging. IEEE Trans. Parallel Distrib. Syst. **33**, 2014 (2022)

46. C. Jailin, S. Roux, Dynamic tomographic reconstruction of deforming volumes. Materials **11** (2018)

47. A. Buljac, C. Jailin, A. Mendoza, J. Neggers, T. Taillandier-Thomas, A. Bouterf, B. Smaniotto, F. Hild, S. Roux, Digital volume correlation: review of progress and challenges. Exp. Mech. **58**, 661 (2018)

48. C. Lo, T. Sano, J.D. Hogan, Microstructural and mechanical characterization of variability in porous advanced ceramics using X-ray computed tomography and digital image correlation. Mater. Charact. **158**, 109929 (2019)

49. E.A. Zwanenburg, M.A. Williams, J.M. Warnett, Review of high-speed imaging with lab-based x-ray computed tomography. Meas. Sci. Technol. **33**, 012003 (2022)

50. Q. Zhu, C. Wang, H. Qin, G. Chen, P. Zhang, Effect of the grain size on the microtensile deformation and fracture behaviors of a nickel-based superalloy via ebsd and in-situ synchrotron radiation X-ray tomography. Mater. Charact. **156**, 109875 (2019)

51. J. Villanova, R. Daudin, P. Lhuissier, D. Jauffrès, S. Lou, C.L. Martin, S. Labouré, R. Tucoulou, G. Martínez-Criado, L. Salvo, Fast in situ 3D nanoimaging: a new tool for dynamic characterization in materials science. Mater. Today **20**, 354 (2017)

52. J. Eller, J. Roth, F. Marone, M. Stampanoni, F.N. Büchi, Operando properties of gas diffusion layers: saturation and liquid permeability. J. Electrochem. Soc. **164**, F115 (2017)

53. H. Xu, F. Marone, S. Nagashima, H. Nguyen, K. Kishita, F.N. Büchi, J. Eller, (Invited) exploring sub-second and sub-micron X-ray tomographic imaging of liquid water in pefc gas diffusion layers. ECS Trans. **92**, 11 (2019)

54. D. Schröder, C.L. Bender, T. Arlt, M. Osenberg, A. Hilger, S. Risse, M. Ballauff, I. Manke, J. Janek, In operando X-ray tomography for next-generation batteries: a systematic approach to monitor reaction product distribution and transport processes. J. Phys. D. Appl. Phys. **49**, 404001 (2016)

55. J.F. Gonzalez, D.A. Antartis, I. Chasiotis, S.J. Dillon, J. Lambros, In situ X-ray micro-CT characterization of chemo-mechanical relaxations during Sn lithiation. J. Power Sources **381**, 181 (2018)

56. D.P. Finegan et al., In-operando high-speed tomography of lithium-ion batteries during thermal runaway. Nat. Commun. **6**, 1 (2015)

57. J. Ohser, D. Dobrovolskij, C. Blankenburg, A. Rack, Time-resolved phase-contrast microtomographic imaging of two-phase solid-liquid flow through porous media. Int. J. Mater. Res. **111**, 86 (2020)

58. V.V. Nikitin, G.A. Dugarov, A.A. Duchkov, M.I. Fokin, A.N. Drobchik, P.D. Shevchenko, F. De Carlo, R. Mokso, Dynamic in-situ imaging of methane hydrate formation and self-preservation in porous media. Mar. Pet. Geol. **115**, 104234 (2020)

59. L. Zhang, K. Ge, J. Wang, J. Zhao, Y. Song, Pore-scale investigation of permeability evolution during hydrate formation using a pore network model based on X-ray CT. Mar. Pet. Geol. **113**, 104157 (2020)

60. P. Perré, D.M. Nguyen, G. Almeida, A macroscopic washburn approach of liquid imbibition in wood derived from X-ray tomography observations. Sci. Rep. **12**, 1750 (2022)

61. L. Holzer, F. Indutnyi, P. Gasser, B. Münch, M. Wegmann, Three-dimensional analysis of porous BaTiO3 ceramics using FIB nanotomography. J. Microsc. **216**, 84 (2004)

62. L. Holzer et al., Fundamental relationships between 3D pore topology, electrolyte conduction and flow properties: towards knowledge-based design of ceramic diaphragms for sensor applications. Mate.r Des. **99**, 314 (2016)

63. L. Holzer, D. Wiedenmann, B. Münch, L. Keller, M. Prestat, P. Gasser, I. Robertson, B. Grobéty, The influence of constrictivity on the effective transport properties of porous layers in electrolysis and fuel cells. J. Mater. Sci. **48**, 2934 (2013)

64. L.M. Keller, L. Holzer, R. Wepf, P. Gasser, 3D geometry and topology of pore pathways in opalinus clay: implications for mass transport. Appl. Clay. Sci. **52**, 85 (2011)

65. L.M. Keller, L. Holzer, R. Wepf, P. Gasser, B. Münch, P. Marschall, On the application of focused ion beam nanotomography in characterizing the 3D pore space geometry of opalinus clay. Phys. Chem. Earth, Parts A/B/C **36**, 1539 (2011)

66. B. Münch, L. Holzer, Contradicting geometrical concepts in pore size analysis attained with electron microscopy and mercury intrusion. J. Am. Ceram. Soc. **91**, 4059 (2008)

67. L.M. Keller, L. Holzer, P. Gasser, R. Erni, M.D. Rossell, Intergranular pore space evolution in mx80 bentonite during a long-term experiment. Appl. Clay Sci. **104** (2015)

68. L.M. Keller, L. Holzer, P. Schuetz, P. Gasser, Pore space relevant for gas permeability in opalinus clay: statistical analysis of homogeneity, percolation, and representative volume element. J. Geophys. Res. Solid Earth **118** (2013)

69. Z. Li, D. Liu, Y. Cai, P.G. Ranjith, Y. Yao, Multi-scale quantitative characterization of 3-D pore-fracture networks in bituminous and anthracite coals using fib-sem tomography and X-ray μ-CT. Fuel **209**, 43 (2017)

70. T. Saif, Q. Lin, A.R. Butcher, B. Bijeljic, M.J. Blunt, Multi-scale multi-dimensional microstructure imaging of oil shale pyrolysis using X-ray micro-tomography. Autom. Ultra-High Resolut. SEM, MAPS Mineral. FIB-SEM, Appl. Energy **202**, 628 (2017)

71. K. Bae, J.W. Kim, J. won Son, T. Lee, S. Kang, F.B. Prinz, J.H. Shim, 3D evaluation of porous zeolite absorbents using FIB-SEM tomography. Int. J. Precis. Eng. Manuf. Green Technol. **5**, 195 (2018)

72. A. Holzinger, G. Neusser, B.J.J. Austen, A. Gamero-Quijano, G. Herzog, D.W.M. Arrigan, A. Ziegler, P. Walther, C. Kranz, Investigation of modified nanopore arrays using FIB/SEM tomography. Faraday Discuss. **210**, 113 (2018)

73. N. Nan, J. Wang, FIB-SEM three-dimensional tomography for characterization of carbon-based materials. Adv. Mater. Sci. Eng. **2019**, 1 (2019)

74. T. Ott, D. Roldán, C. Redenbach, K. Schladitz, M. Godehardt, S. Höhn, Three-dimensional structural comparison of tantalum glancing angle deposition thin films by FIB-SEM. J. Sens. Sens. Syst. **8**, 305 (2019)

75. D.A.M. de Winter, F. Meirer, B.M. Weckhuysen, FIB-SEM tomography probes the mesoscale pore space of an individual catalytic cracking particle. ACS Catal. **6**, 3158 (2016)

76. H. Aslannejad, S.M. Hassanizadeh, A. Raoof, D.A.M. de Winter, N. Tomozeiu, M.T. van Genuchten, Characterizing the hydraulic properties of paper coating layer using FIB-SEM tomography and 3D pore-scale modeling. Chem. Eng. Sci. **160**, 275 (2017)

77. T. Miyaki, Y. Kawasaki, Y. Hosoyamada, T. Amari, M. Kinoshita, H. Matsuda, S. Kakuta, T. Sakai, K. Ichimura, Three-dimensional imaging of podocyte ultrastructure using FE-SEM and FIB-SEM tomography. Cell Tissue Res. **379**, 245 (2020)

78. J.R. Wilson, W. Kobsiriphat, R. Mendoza, H.-Y. Chen, J.M. Hiller, D.J. Miller, K. Thornton, P.W. Voorhees, S.B. Adler, S. a Barnett, Three-dimensional reconstruction of a solid-oxide fuel-cell anode. Nat. Mater. **5**, 541 (2006)

79. N. Shikazono, D. Kanno, K. Matsuzaki, H. Teshima, S. Sumino, N. Kasagi, Numerical assessment of SOFC anode polarization based on three-dimensional model microstructure reconstructed from FIB-SEM images. J. Electrochem. Soc. **157**, B665 (2010)

80. D. Kanno, N. Shikazono, N. Takagi, K. Matsuzaki, N. Kasagi, Evaluation of SOFC anode polarization simulation using three-dimensional microstructures reconstructed by FIB tomography. Electrochim. Acta **56**, 4015 (2011)

81. H. Iwai et al., Quantification of SOFC anode microstructure based on dual beam FIB-SEM technique. J. Power Sources **195**, 955 (2010)

82. O. Pecho, A. Mai, B. Münch, T. Hocker, R. Flatt, L. Holzer, 3D microstructure effects in Ni-YSZ anodes: influence of TPB lengths on the electrochemical performance. Materials **8**, 7129 (2015)

83. O. Pecho, O. Stenzel, B. Iwanschitz, P. Gasser, M. Neumann, V. Schmidt, M. Prestat, T. Hocker, R. Flatt, L. Holzer, 3D microstructure effects in Ni-YSZ anodes: prediction of effective transport properties and optimization of redox stability. Materials **8**, 5554 (2015)

84. L. Holzer, B. Iwanschitz, Th. Hocker, L. Keller, O. Pecho, G. Sartoris, Ph. Gasser, B. Muench, Redox cycling of Ni–YSZ anodes for solid oxide fuel cells: influence of tortuosity, constriction and percolation factors on the effective transport properties. J. Power Sources **242**, 179 (2013)

85. N. Vivet, S. Chupin, E. Estrade, a. Richard, S. Bonnamy, D. Rochais, E. Bruneton, Effect of Ni content in SOFC Ni-YSZ cermets: a three-dimensional study by FIB-SEM tomography, J. Power Sources **196**, 9989 (2011)

86. N. Vivet, S. Chupin, E. Estrade, T. Piquero, P.L. Pommier, D. Rochais, E. Bruneton, 3D microstructural characterization of a solid oxide fuel cell anode reconstructed by focused ion beam tomography. J. Power Sources **196**, 7541 (2011)

87. A. Zekri, M. Knipper, J. Parisi, T. Plaggenborg, Microstructure degradation of Ni/CGO anodes for solid oxide fuel cells after long operation time using 3d reconstructions by FIB tomography. Phys. Chem. Chem. Phys. **19**, 13767 (2017)

88. M. Meffert, F. Wankmüller, H. Störmer, A. Weber, P. Lupetin, E. Ivers-Tiffée, D. Gerthsen, Optimization of material contrast for efficient FIB-SEM tomography of solid oxide fuel cells. Fuel Cells **20**, 580 (2020)

89. Z. Liu, Y.K. Chen-Wiegart, J. Wang, S.A. Barnett, K.T. Faber, Three-phase 3D reconstruction of a LiCoO2 cathode via FIB-SEM tomography. Microsc. Microanal. **22**, 140 (2016)

90. H. Liu, J.M. Foster, A. Gully, S. Krachkovskiy, M. Jiang, Y. Wu, X. Yang, B. Protas, G.R. Goward, G.A. Botton, Three-dimensional investigation of cycling-induced microstructural changes in lithium-ion battery cathodes using focused ion beam/scanning electron microscopy. J. Power Sources **306**, 300 (2016)

91. A. Etiemble, A. Tranchot, T. Douillard, H. Idrissi, E. Maire, L. Roué, Evolution of the 3D microstructure of a si-based electrode for Li-ion batteries investigated by FIB/SEM tomography. J. Electrochem. Soc. **163**, A1550 (2016)

92. L. Almar, J. Joos, A. Weber, E. Ivers-Tiffée, Microstructural feature analysis of commercial Li-ion battery cathodes by focused ion beam tomography. J. Power Sources **427**, 1 (2019)

93. A.C. Wagner, N. Bohn, H. Geßwein, M. Neumann, M. Osenberg, A. Hilger, I. Manke, V. Schmidt, J.R. Binder, Hierarchical structuring of NMC111-cathode materials in lithium-ion batteries: an in-depth study on the influence of primary and secondary particle sizes on electrochemical performance. ACS Appl. Energy Mater. **3**, 12565 (2020)

94. L. Zielke, T. Hutzenlaub, D.R. Wheeler, C.-W. Chao, I. Manke, A. Hilger, N. Paust, R. Zengerle, S. Thiele, Three-phase multiscale modeling of a LiCoO 2 cathode: combining the advantages of FIB-SEM imaging and X-ray tomography. Adv. Energy Mater. **5**, 1401612 (2015)

95. A. Kruk, G. Cempura, S. Lech, A. Czyrska -Filemonowicz, Stem-EDX and FIB-SEM tomography of ALLVAC 718Plus superalloy. Arch. Metall. Mater. **61**, 535 (2016)

96. S.S. Singh, J.J. Loza, A.P. Merkle, N. Chawla, Three dimensional microstructural characterization of nanoscale precipitates in AA7075-T651 by focused ion beam (FIB) tomography. Mater. Charact. **118**, 102 (2016)

97. K. Jahns, U. Krupp, G. Sundell, C. Geers, Formation of corrosion pockets in FeNiCrAl at high temperatures investigated by 3D FIB-SEM tomography. Mater. Corros. **71**, 1774 (2020)

98. P.G. Kotula, G.S. Rohrer, M.P. Marsh, Focused ion beam and scanning electron microscopy for 3D materials characterization. MRS Bull. **39**, 361 (2014)

99. O. Furat, L. Petrich, D.P. Finegan, D. Diercks, F. Usseglio-Viretta, K. Smith, V. Schmidt, Artificial generation of representative single Li-ion electrode particle architectures from microscopy data. NPJ Comput. Mater. **7**, 105 (2021)

100. O. Furat, D.P. Finegan, D. Diercks, F. Usseglio-Viretta, K. Smith, V. Schmidt, Mapping the architecture of single lithium ion electrode particles in 3D, using electron backscatter diffraction and machine learning segmentation. J. Power Sources **483**, 229148 (2021)

101. A. Quinn, H. Moutinho, F. Usseglio-Viretta, A. Verma, K. Smith, M. Keyser, D.P. Finegan, Electron backscatter diffraction for investigating Lithium-ion electrode particle architectures. Cell Rep. Phys. Sci. **1**, 100137 (2020)

102. L. Holzer, P.H. Gasser, A. Kaech, M. Wegmann, A. Zingg, R. Wepf, B. Muench, Cryo-FIB-nanotomography for quantitative analysis of particle structures in cement suspensions. J. Microsc. **227**, 216 (2007)

103. A. Zingg, L. Holzer, A. Kaech, F. Winnefeld, J. Pakusch, S. Becker, L. Gauckler, The microstructure of dispersed and non-dispersed fresh cement pastes—new insight by cryo-microscopy. Cem. Concr. Res. **38**, 522 (2008)

104. L. Holzer, B. Münch, Toward reproducible three-dimensional microstructure analysis of granular materials and complex suspensions. Microsc. Microanal. **15**, 130 (2009)

105. L. Holzer, B. Münch, M. Rizzi, R. Wepf, P. Marschall, T. Graule, 3D-microstructure analysis of hydrated bentonite with cryo-stabilized pore water. Appl. Clay Sci. **47**, 330 (2010)

106. L. Holzer, M. Cantoni, in *Review of FIB-Tomography*, in *Nanofabrication Using Focused Ion and Electron Beams: Principles and Applications*, ed. by I. Utke, S. Moshkalev, Ph. Russell (Oxford University Press, New York, 2011), pp. 410–435

107. M. Cantoni, L. Holzer, Advances in 3D focused ion beam tomography. MRS Bull **39**, 354 (2014)

108. S.N. Monteiro, S. Paciornik, From historical backgrounds to recent advances in 3D characterization of materials: an overview. JOM **69**, 84 (2017)

109. M.P. Echlin, T.L. Burnett, A.T. Polonsky, T.M. Pollock, P.J. Withers, Serial sectioning in the SEM for three dimensional materials science. Curr. Opin. Solid State Mater. Sci. **24**, 100817 (2020)
110. T.L. Burnett, R. Kelley, B. Winiarski, L. Contreras, M. Daly, A. Gholinia, M.G. Burke, P.J. Withers, Large volume serial section tomography by Xe plasma FIB dual beam microscopy. Ultramicroscopy **161**, 119 (2016)
111. Y. Zhang, C. Kong, R.S. Davidsen, G. Scardera, L. Duan, K.T. Khoo, D.N.R. Payne, B. Hoex, M. Abbott, 3D characterisation using plasma FIB-SEM: a large-area tomography technique for complex surfaces like black silicon. Ultramicroscopy **218**, 113084 (2020)
112. N. Bassim, K. Scott, L.A. Giannuzzi, Recent advances in focused ion beam technology and applications. MRS Bull. **39**, 317 (2014)
113. M.P. Echlin, M. Straw, S. Randolph, J. Filevich, T.M. Pollock, The TriBeam system: femtosecond laser ablation in situ SEM. Mater. Charact. **100**, 1 (2015)
114. Y. Zhang, C. Kong, G. Scardera, M. Abbott, D.N.R. Payne, B. Hoex, Large Volume tomography using plasma FIB-SEM: a comprehensive case study on black silicon. Ultramicroscopy **233**, 113458 (2022)
115. B. Winiarski, A. Gholinia, K. Mingard, M. Gee, G.E. Thompson, P.J. Withers, Broad ion beam serial section tomography. Ultramicroscopy **172**, 52 (2017)
116. A. Gholinia, M.E. Curd, E. Bousser, K. Taylor, T. Hosman, S. Coyle, M.H. Shearer, J. Hunt, P.J. Withers, Coupled broad ion beam-scanning electron microscopy (BIB–SEM) for polishing and three dimensional (3D) serial section tomography (SST). Ultramicroscopy **214**, 112989 (2020)
117. S.J. Randolph, J. Filevich, A. Botman, R. Gannon, C. Rue, M. Straw, In situ femtosecond pulse laser ablation for large volume 3d analysis in scanning electron microscope systems. J. Vac. Sci. Technol. B **36**, 06JB01 (2018)
118. S. Randolph, R. Geurts, J. Wang, B. Winiarski, C. Rue, Femtosecond laser-enabled tribeam as a platform for analysis of thermally- and charge-sensitive materials. Microsc. Microanal. **25**, 352 (2019)
119. W. Denk, H. Horstmann, Serial block-face scanning electron microscopy to reconstruct three-dimensional tissue nanostructure. PLoS Biol. **2**, e329 (2004)
120. D.J. Rowenhorst, L. Nguyen, A.D. Murphy-Leonard, R.W. Fonda, Characterization of microstructure in additively manufactured 316l using automated serial sectioning. Curr. Opin. Solid State Mater. Sci. **24**, 100819 (2020)
121. R. Mahbub, T. Hsu, W.K. Epting, N.T. Nuhfer, G.A. Hackett, H. Abernathy, A.D. Rollett, M. De Graef, S. Litster, P.A. Salvador, A method for quantitative 3D mesoscale analysis of solid oxide fuel cell microstructures using Xe-plasma focused ion beam (PFIB) coupled with SEM. ECS Trans. **78**, 2159 (2017)
122. R. Erni, M.D. Rossell, C. Kisielowski, U. Dahmen, Atomic-resolution imaging with a sub-50-Pm electron probe. Phys. Rev. Lett. **102**, 096101 (2009)
123. P. Ercius, O. Alaidi, M.J. Rames, G. Ren, Electron tomography: a three-dimensional analytic tool for hard and soft materials research. Adv. Mater. **27**, 5638 (2015)
124. H. Song, Y. Yang, J. Geng, Z. Gu, J. Zou, C. Yu, Electron tomography: a unique tool solving intricate hollow nanostructures. Adv. Mater. **31**, 1 (2019)
125. R. Hovden, D. Muller, Electron tomography for functional nanomaterials. ArXiv 1 (2020)
126. A.V. Ceguerra, R.K.W. Marceau, Atom probe tomography of aluminium alloys: a systematic meta-analysis review of 2018. Metals (Basel) **9**, 1 (2019)
127. K. Eder, I. McCarroll, A. La Fontaine, J.M. Cairney, Nanoscale analysis of corrosion products: a review of the application of atom probe and complementary microscopy techniques. Jom **70**, 1744 (2018)
128. P. Dumitraschkewitz, S.S.A. Gerstl, P.J. Uggowitzer, J.F. Löffler, S. Pogatscher, Atom probe tomography study of as-quenched Al–Mg–Si alloys. Adv. Eng. Mater. **19**, 1 (2017)
129. I.E. McCarroll, P.A.J. Bagot, A. Devaraj, D.E. Perea, J.M. Cairney, New frontiers in atom probe tomography: a review of research enabled by cryo and/or vacuum transfer systems. Mater. Today Adv. **7**, 100090 (2020)

130. D.E. Perea, D.K. Schreiber, J.V. Ryan, M.G. Wirth, L. Deng, X. Lu, J. Du, J.D. Vienna, Tomographic mapping of the nanoscale water-filled pore structure in corroded borosilicate glass. Npj Mater. Degrad. **4**, 1 (2020)

131. P. Paul-Gilloteaux, X. Heiligenstein, M. Belle, M.-C. Domart, B. Larijani, L. Collinson, G. Raposo, J. Salamero, EC-CLEM: flexible multidimensional registration software for correlative microscopies. Nat. Methods **14**, 102 (2017)

132. S. Handschuh, N. Baeumler, T. Schwaha, B. Ruthensteiner, A correlative approach for combining MicroCT, light and transmission electron microscopy in a single 3D scenario. Front. Zool. **10**, 44 (2013)

133. F.G.A. Faas, M.C. Avramut, B.M. van den Berg, A.M. Mommaas, A.J. Koster, R.B.G. Ravelli, Virtual nanoscopy: generation of ultra-large high resolution electron microscopy maps. J. Cell Biol. **198**, 457 (2012)

134. D.R. Glenn, H. Zhang, N. Kasthuri, R. Schalek, P.K. Lo, A.S. Trifonov, H. Park, J.W. Lichtman, R.L. Walsworth, Correlative light and electron microscopy using cathodoluminescence from nanoparticles with distinguishable colours. Sci. Rep. **2**, 865 (2012)

135. J. Caplan, M. Niethammer, R.M. Taylor, K.J. Czymmek, The power of correlative microscopy: multi-modal, multi-scale, multi-dimensional. Curr. Opin. Struct. Biol. **21**, 686 (2011)

136. T.L. Burnett et al., Correlative tomography. Sci. Rep. **4**, 4711 (2015)

137. T.L. Burnett, P.J. Withers, Completing the picture through correlative characterization. Nat. Mater. **18**, 1041 (2019)

138. R.S. Bradley, P.J. Withers, Correlative multiscale tomography of biological materials. MRS Bull. **41**, 549 (2016)

139. P.R. Shearing, N.P. Brandon, J. Gelb, R. Bradley, P.J. Withers, A.J. Marquis, S. Cooper, S.J. Harris, Multi length scale microstructural investigations of a commercially available Li-ion battery electrode. J. Electrochem. Soc. **159**, A1023 (2012)

140. A. Kwiatkowski da Silva, G. Leyson, M. Kuzmina, D. Ponge, M. Herbig, S. Sandlöbes, B. Gault, J. Neugebauer, D. Raabe, Confined chemical and structural states at dislocations in Fe-9wt%Mn steels: a correlative TEM-atom probe study combined with multiscale modelling. Acta Mater. **124**, 305 (2017)

141. Y. Fam, T.L. Sheppard, A. Diaz, T. Scherer, M. Holler, W. Wang, D. Wang, P. Brenner, A. Wittstock, J.-D. Grunwaldt, Correlative multiscale 3D imaging of a hierarchical nanoporous gold catalyst by electron. Ion X-Ray Nanotomography, ChemCatChem **10**, 2858 (2018)

142. L.M. Keller, L. Holzer, Image-based upscaling of permeability in opalinus clay. J. Geophys. Res. Solid Earth **123**, 285 (2018)

143. L.M. Keller, P. Schuetz, R. Erni, M.D. Rossell, F. Lucas, P. Gasser, L. Holzer, Characterization of multi-scale microstructural features in opalinus clay. Microporous Mesoporous Mater. **170**, 83 (2013)

144. S.N. Chiu, D. Stoyan, W. Kendall, J. Mecke, *Stochastic Geometry and Its Applications*, 3rd edn. (Wiley, Chichester, UK, 2013)

145. J. Ohser, K. Schladitz, *3D Images of Materials Structures: Processing and Analysis* (Wiley, Weinheim, Germany, 2009)

146. J. Serra, *Image Analysis and Mathematical Morphology* (Academic Press Ltd., London, 1982)

147. S.J. Cooper, A. Bertei, P.R. Shearing, J.A. Kilner, N.P. Brandon, TauFactor: an open-source application for calculating tortuosity factors from tomographic data. SoftwareX **5**, 203 (2016)

148. M. Ebner, V. Wood, Tool for tortuosity estimation in lithium ion battery porous electrodes. J. Electrochem. Soc. **162**, A3064 (2015)

149. B. Münch, *Empa Bundle of ImageJ Plugins for Image Analysis (EBIPIA)*, http://Wiki.Imagej. Net/Xlib

150. S. Barman, D. Bolin, C. Fager, T. Gebäck, N. Lorén, E. Olaaon, H. Rootzén, A. Särkkä, *Mist—A Program Package for Visualization and Characterization of 3D Geometries*

151. S. Barman, H. Rootzén, D. Bolin, Prediction of diffusive transport through polymer films from characteristics of the pore geometry. AIChE J. **65**, 446 (2019)

152. M. Matyka, Z. Koza, How to calculate tortuosity easily? AIP Conf. Proc. **1453**, 17 (2012)

153. J. Schindelin et al., Fiji: an open-source platform for biological-image analysis. Nat. Methods **9**, 676 (2012)
154. W. Van Aarle, W.J. Palenstijn, J. De Beenhouwer, T. Altantzis, S. Bals, K.J. Batenburg, J. Sijbers, The ASTRA toolbox: a platform for advanced algorithm development in electron tomography. Ultramicroscopy **157**, 157 (2015)
155. J.C. Russ, *The Image Processing Handbook, Sixth Edition - CRC Press Book* (CRC Press, Boca Raton, USA, 2011)
156. S. Schlüter, A. Sheppard, K. Brown, D. Wildenschild, Image processing of multiphase images obtained via X-Ray microtomography: a review. Water Resour. Res. **50**, 3615 (2014)
157. S. Berg et al., Ilastik: interactive machine learning for (bio)image analysis. Nat. Methods **16**, 1226 (2019)
158. I. Arganda-Carreras, V. Kaynig, C. Rueden, K.W. Eliceiri, J. Schindelin, A. Cardona, H. Sebastian Seung, Trainable weka segmentation: a machine learning tool for microscopy pixel classification. Bioinformatics **33**, 2424 (2017)
159. O. Furat, M. Wang, M. Neumann, L. Petrich, M. Weber, C.E. Krill, V. Schmidt, Machine learning techniques for the segmentation of tomographic image data of functional materials. Front. Mater. **6** (2019)
160. C. Fend, A. Moghiseh, C. Redenbach, K. Schladitz, Reconstruction of highly porous structures from FIB-SEM using a deep neural network trained on synthetic images. J. Microsc. **281**, 16 (2021)
161. A. Buades, B. Coll, J.-M. Morel, *A Non-Local Algorithm for Image Denoising*, in *IEEE Computer Society Conference on Computer Vision and Pattern Recognition (CVPR'05)*, vol 2 (IEEE, 2005), pp. 60–65
162. B. Tjaden, D.J.L. Brett, P.R. Shearing, Tortuosity in electrochemical devices: a review of calculation approaches. Int. Mater. Rev. **63**, 47 (2018)
163. J. Fu, H.R. Thomas, C. Li, Tortuosity of porous media: image analysis and physical simulation. Earth Sci. Rev. 1 (2020)
164. O. Stenzel, O. Pecho, L. Holzer, M. Neumann, V. Schmidt, Predicting effective conductivities based on geometric microstructure characteristics. AIChE J. **62**, 1834 (2016)
165. M. Neumann, O. Stenzel, F. Willot, L. Holzer, V. Schmidt, Quantifying the influence of microstructure on effective conductivity and permeability: virtual materials testing. Int. J. Solids Struct. **184**, 211 (2020)
166. W.B. Lindquist, S.-M. Lee, D.A. Coker, K.W. Jones, P. Spanne, Medial axis analysis of void structure in three-dimensional tomographic images of porous media. J. Geophys. Res. Solid Earth **101**, 8297 (1996)
167. J. Vicente, F. Topin, J.V. Daurelle, Open celled material structural properties measurement: from morphology to transport properties. Mater. Trans. **47**, 2195 (2006)
168. P.S. Jørgensen, K.V. Hansen, R. Larsen, J.R. Bowen, Geometrical characterization of interconnected phase networks in three dimensions. J. Microsc. **244**, 45 (2011)
169. T. Hamann, L. Zhang, Y. Gong, G. Godbey, J. Gritton, D. McOwen, G. Hitz, and E. Wachsman, The effects of constriction factor and geometric tortuosity on li-ion transport in porous solid-state Li-ion electrolytes. Adv. Funct. Mater. **30** (2020)
170. W. Sobieski, The use of path tracking method for determining the tortuosity field in a porous bed. Granul. Matter. **18**, 1 (2016)
171. W. Sobieski, M. Matyka, J. Gołembiewski, S. Lipiński, The path tracking method as an alternative for tortuosity determination in granular beds. Granul. Matter. **20** (2018)
172. W. Sobieski, Numerical investigations of tortuosity in randomly generated pore structures. Math. Comput. Simul. **166**, 1 (2019)
173. T.G. Tranter, M.D.R. Kok, M. Lam, J.T. Gostick, Pytrax: a simple and efficient random walk implementation for calculating the directional tortuosity of images. SoftwareX **10**, 100277 (2019)
174. J. Latt et al., Palabos: parallel lattice boltzmann solver. Comput. Math. Appl. **81**, 334 (2021)
175. J. Van Brakel, P.M. Heertjes, Analysis of diffusion in macroporous media in terms of a porosity, a tortuosity and a constrictivity factor. Int. J. Heat Mass Transf. 1093 (1974)

176. C.F. Berg, Permeability description by characteristic length, tortuosity, constriction and porosity. Transp. Porous Media **103**, 381 (2014)
177. C.F. Berg, Re-examining Archie's law: conductance description by tortuosity and constriction. Phys. Rev. E **86**, 046314 (2012)
178. J. Hoogschagen, Diffusion in porous catalysts and adsorbents. Ind. Eng. Chem. **47**, 906 (1955)
179. P.C. Carman, Fluid flow through granular beds. Chem. Eng. Res. Des. **75**, S32 (1997)
180. A. Koponen, M. Kataja, J. Timonen, Permeability and effective porosity of porous media. Phys. Rev. E **56**, 3319 (1997)
181. A. Koponen, M. Kataja, J. Timonen, Tortuous flow in porous media. Phys. Rev. E **54**, 406 (1996)
182. M.A. Knackstedt, X. Zhang, Direct evaluation of length scales and structural parameters associated with flow in porous media. Phys. Rev. E **50**, 2134 (1994)
183. M. Matyka, A. Khalili, Z. Koza, Tortuosity-porosity relation in porous media flow. Phys. Rev. E **78**, 026306 (2008)
184. Z. Koza, M. Matyka, A. Khalili, Finite-size anisotropy in statistically uniform porous media. Phys. Rev. E **79**, 066306 (2009)
185. R. Nemati, J. Rahbar Shahrouzi, R. Alizadeh, A stochastic approach for predicting tortuosity in porous media via pore network modeling, Comput. Geotech. **120**, 103406 (2020)
186. A. Duda, Z. Koza, and M. Matyka, Hydraulic tortuosity in arbitrary porous media flow. Phys. Rev. E Stat. Nonlin. Soft Matter. Phys. **84** (2011)
187. R.B. Bird, W.E. Steward, E.N. Lightfood, *Transport Phenomena, Second Edi* (Wiley, New York, 2007)
188. M. Sahimi, *Flow and Transport in Porous Media and Fractured Rock: From Classical Methods to Modern Approaches* (Wiley, 2011)
189. J. Bear, *Modeling Phenomena of Flow and Transport in Porous Media*, vol. 1 (Springer International Publishing, Cham, 2018)
190. N. Saxena, R. Hofmann, F.O. Alpak, S. Berg, J. Dietderich, U. Agarwal, K. Tandon, S. Hunter, J. Freeman, O.B. Wilson, References and benchmarks for pore-scale flow simulated using micro-CT images of porous media and digital rocks. Adv. Water Resour. **109**, 211 (2017)
191. J. Su, L. Wang, Z. Gu, Y. Zhang, C. Chen, *Advances in Pore-Scale Simulation of Oil Reservoirs*. Energies (Basel) **11**, (2018)
192. J. He, Y. Ju, L. Lammers, K. Kulasinski, L. Zheng, Tortuosity of kerogen pore structure to gas diffusion at molecular- and nano-scales: a molecular dynamics simulation. Chem. Eng. Sci. **215**, 115460 (2020)
193. W. Wang, D. Fan, G. Sheng, Z. Chen, Y. Su, A review of analytical and semi-analytical fluid flow models for ultra-tight hydrocarbon reservoirs. Fuel **256**, 115737 (2019)
194. D. Müter, H.O. Sørensen, H. Bock, S.L.S. Stipp, Particle diffusion in complex nanoscale pore networks. J. Phys. Chem. C **119**, 10329 (2015)
195. U. Tallarek, D. Hlushkou, J. Rybka, A. Höltzel, Multiscale simulation of diffusion in porous media: from interfacial dynamics to hierarchical porosity. J. Phys. Chem. C **123**, 15099 (2019)
196. B. Ghanbarian, Scale dependence of tortuosity and diffusion: finite-size scaling analysis. J. Contam. Hydrol. **245**, 103953 (2022)
197. E.M. Ryan, P.P. Mukherjee, Mesoscale modeling in electrochemical devices—a critical perspective. Prog. Energy Combust. Sci. **71**, 118 (2019)
198. F.L.E. Usseglio-Viretta et al., Resolving the discrepancy in tortuosity factor estimation for li-ion battery electrodes through micro-macro modeling and experiment. J. Electrochem. Soc. **165**, A3403 (2018)
199. X. Lu et al., 3D microstructure design of lithium-ion battery electrodes assisted by X-ray nano-computed tomography and modelling. Nat. Commun. **11**, 1 (2020)
200. J. Le Houx, D. Kramer, Physics based modelling of porous lithium ion battery electrodes—a review. Energy Rep. **6**, 1 (2020)
201. D. Zhang, A. Forner-Cuenca, O.O. Taiwo, V. Yufit, F.R. Brushett, N.P. Brandon, S. Gu, Q. Cai, Understanding the role of the porous electrode microstructure in redox flow battery performance using an experimentally validated 3d pore-scale lattice boltzmann model. J. Power Sources **447**, 227249 (2020)

202. L. Chen, A. He, J. Zhao, Q. Kang, Z.-Y. Li, J. Carmeliet, N. Shikazono, W.-Q. Tao, Pore-scale modeling of complex transport phenomena in porous media. Prog. Energy Combust. Sci. **88**, 100968 (2022)

203. H. Tao, G. Chen, C. Lian, H. Liu, M. Coppens, Multiscale modeling of ion transport in porous electrodes, AIChE J. **68** (2022)

204. K.N. Grew, W.K.S. Chiu, A review of modeling and simulation techniques across the length scales for the solid oxide fuel cell. J. Power Sources **199**, 1 (2012)

205. J. Hanna, W.Y. Lee, Y. Shi, A.F. Ghoniem, Fundamentals of electro- and thermochemistry in the anode of solid-oxide fuel cells with hydrocarbon and syngas fuels. Prog. Energy Combust. Sci. **40**, 74 (2014)

206. S. Dierickx, J. Joos, A. Weber, E. Ivers-Tiffée, Advanced impedance modelling of Ni/8YSZ cermet anodes. Electrochim. Acta **265**, 736 (2018)

207. B. Timurkutluk, M.D. Mat, A review on micro-level modeling of solid oxide fuel cells. Int. J. Hydrogen Energy **41**, 9968 (2016)

208. A.Z. Weber et al., A critical review of modeling transport phenomena in polymer-electrolyte fuel cells. J. Electrochem. Soc. **161**, F1254 (2014)

209. I.V. Zenyuk, E. Medici, J. Allen, A.Z. Weber, Coupling continuum and pore-network models for polymer-electrolyte fuel cells. Int. J. Hydrogen Energy **40**, 16831 (2015)

210. X. Liu, F. Peng, G. Lou, Z. Wen, Liquid water transport characteristics of porous diffusion media in polymer electrolyte membrane fuel cells: a review. J. Power Sources **299**, 85 (2015)

211. G. Matheron, *Random Sets and Integral Geometry* (Wiley, New York, 1975)

212. D. Jeulin, *Morphological Models of Random Structures* (Springer, Cham, 2021)

213. C. Lantuéjoul, *Geostatistical Simulation: Models and Algorithms* (Springer, Berlin, 2013)

214. V. Schmidt, *Stochastic Geometry, Spatial Statistics and Random Fields* (Springer, Cham, 2014)

215. S. Bargmann, B. Klusemann, J. Markmann, J.E. Schnabel, K. Schneider, C. Soyarslan, J. Wilmers, Generation of 3D representative volume elements for heterogeneous materials: a review. Prog. Mater. Sci. **96**, 322 (2018)

216. B. Sheikh, A. Pak, Numerical investigation of the effects of porosity and tortuosity on soil permeability using coupled three-dimensional discrete-element method and lattice boltzmann method. Phys. Rev. E Stat. Nonlin. Soft Matter. Phys. **91**, 1 (2015)

217. C. Redenbach, Microstructure models for cellular materials. Comput. Mater. Sci. **44**, 1397 (2009)

218. M. Geißendörfer, A. Liebscher, C. Proppe, C. Redenbach, D. Schwarzer, Stochastic multiscale modeling of metal foams. Probab. Eng. Mech. **37**, 132 (2014)

219. S. Föhst, S. Osterroth, F. Arnold, C. Redenbach, Influence of geometry modifications on the permeability of open-cell foams. AIChE J. **68** (2022)

220. D. Westhoff, J. Skibinski, O. Šedivý, B. Wysocki, T. Wejrzanowski, V. Schmidt, Investigation of the relationship between morphology and permeability for open-cell foams using virtual materials testing. Mater. Des. **147**, 1 (2018)

221. R. Thiedmann, C. Hartnig, I. Manke, V. Schmidt, W. Lehnert, Local structural characteristics of pore space in gdls of pem fuel cells based on geometric 3D graphs. J. Electrochem. Soc. **156**, B1339 (2009)

222. Z. Tayarani-Yoosefabadi, D. Harvey, J. Bellerive, E. Kjeang, Stochastic microstructural modeling of fuel cell gas diffusion layers and numerical determination of transport properties in different liquid water saturation levels. J. Power Sources **303**, 208 (2016)

223. S. Hein, J. Feinauer, D. Westhoff, I. Manke, V. Schmidt, A. Latz, Stochastic microstructure modeling and electrochemical simulation of lithium-ion cell anodes in 3D. J. Power Sources **336**, 161 (2016)

224. D. Westhoff, J. Feinauer, K. Kuchler, T. Mitsch, I. Manke, S. Hein, A. Latz, V. Schmidt, Parametric stochastic 3D model for the microstructure of anodes in lithium-ion power cells. Comput. Mater. Sci. **126**, 453 (2017)

225. T. Hofmann, D. Westhoff, J. Feinauer, H. Andrä, J. Zausch, V. Schmidt, R. Müller, Electro-chemo-mechanical simulation for lithium ion batteries across the scales. Int. J. Solids Struct. **184**, 24 (2020)

226. D. Westhoff, I. Manke, V. Schmidt, Generation of virtual lithium-ion battery electrode microstructures based on spatial stochastic modeling. Comput. Mater. Sci. **151**, 53 (2018)

227. B. Prifling, D. Westhoff, D. Schmidt, H. Markötter, I. Manke, V. Knoblauch, V. Schmidt, Parametric microstructure modeling of compressed cathode materials for Li-Ion batteries. Comput. Mater. Sci. **169**, 109083 (2019)

228. B. Prifling, M. Ademmer, F. Single, O. Benevolenski, A. Hilger, M. Osenberg, I. Manke, V. Schmidt, Stochastic 3D microstructure modeling of anodes in lithium-ion batteries with a particular focus on local heterogeneity. Comput. Mater. Sci. **192**, 110354 (2021)

229. J. Feinauer, T. Brereton, A. Spettl, M. Weber, I. Manke, V. Schmidt, Stochastic 3D modeling of the microstructure of lithium-ion battery anodes via gaussian random fields on the sphere. Comput. Mater. Sci. **109**, 137 (2015)

230. H. Xu, J. Zhu, D.P. Finegan, H. Zhao, X. Lu, W. Li, N. Hoffman, A. Bertei, P. Shearing, M.Z. Bazant, Guiding the design of heterogeneous electrode microstructures for Li-Ion batteries: microscopic imaging. Predictive Model. Mach. Learn. Adv. Energy Mater. **11**, 2003908 (2021)

231. Y. Suzue, N. Shikazono, N. Kasagi, Micro modeling of solid oxide fuel cell anode based on stochastic reconstruction. J. Power Sources **184**, 52 (2008)

232. M. Neumann, J. Staněk, O.M. Pecho, L. Holzer, V. Beneš, V. Schmidt, Stochastic 3D modeling of complex three-phase microstructures in SOFC-electrodes with completely connected phases. Comput. Mater. Sci. **118**, 353 (2016)

233. H. Moussaoui, J. Laurencin, Y. Gavet, G. Delette, M. Hubert, P. Cloetens, T. Le Bihan, J. Debayle, Stochastic geometrical modeling of solid oxide cells electrodes validated on 3D reconstructions. Comput. Mater. Sci. **143**, 262 (2018)

234. B. Abdallah, F. Willot, D. Jeulin, Morphological modelling of three-phase microstructures of anode layers using SEM images. J. Microsc. **263**, 51 (2016)

235. M. Neumann, B. Abdallah, L. Holzer, F. Willot, V. Schmidt, Stochastic 3D modeling of three-phase microstructures for predicting transport properties: a case study. Transp. Porous. Media **128** (2019)

236. L.L. Schepp et al., Digital rock physics and laboratory considerations on a high-porosity volcanic rock. Sci. Rep. **10**, 1 (2020)

237. M. Prodanovic, M. Esteva, M. Hanlon, G. Nanda, P. Agarwal, *Digital Rocks Portal: A Repository for Porous Media Images* (2015)

238. B. Prifling, M. Röding, P. Townsend, M. Neumann, V. Schmidt, Large-scale statistical learning for mass transport prediction in porous materials using 90,000 artificially generated microstructures, Submitted (2022)

Chapter 5
Towards a Quantitative Understanding of Microstructure-Property Relationships

Abstract 100 years ago, the concept of tortuosity was introduced by Kozeny in order to express the limiting influence of the microstructure on porous media flow. It was also recognized that transport is hindered by other microstructure features such as pore volume fraction, narrow bottlenecks, and viscous drag at the pore surface. The ground-breaking work of Kozeny and Carman makes it possible to predict the macroscopic flow properties (i.e., permeability) based on the knowledge of the relevant microstructure characteristics. However, Kozeny and Carman did not have access to tomography and 3D image analysis techniques, as it is the case nowadays. So, their descriptions were developed by considering simplified models of porous media such as parallel tubes and sphere packings. This simplified setting clearly limits the prediction power of the Carman-Kozeny equations, especially for materials with complex microstructures. Since the ground-breaking work of Kozeny and Carman many attempts were undertaken to improve the prediction power of quantitative expressions that describe the relationship between microstructure characteristics (i.e., tortuosity τ, constrictivity β, porosity ε, hydraulic radius r_h) and effective transport properties (i.e., conductivity σ_{eff}, diffusivity D_{eff}, permeability κ). Due to the ongoing progress in tomography, 3D image-processing, stochastic geometry and numerical simulation, new possibilities arise for better descriptions of the relevant microstructure characteristics, which also leads to mathematical expressions with higher prediction power. In this chapter, the 100-years evolution of quantitative expressions describing the micro–macro relationships in porous media is carefully reviewed,—first, for the case of conduction and diffusion,—and second, for flow and permeability.

The following expressions are the once with the highest prediction power:

$$\sigma_{eff}\left(or\, D_{eff}\right) = \varepsilon^{1.15}\beta^{0.37}/\tau_{dir_{geodesic}}^{4.39},$$

for conduction and diffusion, and

$$\kappa_I = 0.54\left(\frac{\varepsilon}{S_V}\right)^2 \frac{\varepsilon^{3.56}\beta^{0.78}}{\tau_{dir\_geodesic}^{1.67}},$$

L. Holzer et al., *Tortuosity and Microstructure Effects in Porous Media*,
Springer Series in Materials Science 333,
https://doi.org/10.1007/978-3-031-30477-4_5

$$\kappa_{II} = \frac{(0.94 r_{min} + 0.06 r_{max})^2}{8} \frac{\varepsilon^{2.14}}{\tau_{dir\_geodesic}^{2.44}},$$

both, for permeability in porous media.

5.1 Introduction

In this chapter, empirical relationships between morphological characteristics (porosity, tortuosity, constrictivity, hydraulic radius) and macroscopic transport properties (effective conductivity, effective diffusivity, and permeability) are described. Based on the rapid progress of analytical techniques (i.e., 3D imaging, image processing, stochastic simulation, numerical modeling, cloud computing) the prediction power of such equations has improved considerably over the last decade. The newest formulations are now capable to predict micro–macro relationships for many different types of materials and microstructures (e.g., granular, fibrous, cellular, and platy microstructures) in a reliable way. Consequently, these empirical relationships do have some general meaning although they are not derived from a rigorous theoretical basis.

It must be emphasized that microstructure effects limiting the transport in porous media can be investigated in different ways and with different methodologies. In this context, transport simulations based on 3D microstructure models are particularly well suited to predict effective transport properties of porous media. However, the transport simulations themselves cannot replace the valuable information from micro–macro relationships. For example, in materials engineering these empirical relationships provide a unique basis for controlled microstructure optimization and associated materials design, which cannot be replaced by transport simulations. In contrast, for settings with complex transport mechanisms, the concept of tortuosity typically comes to its limits and sophisticated transport simulations are better suited to study the properties of porous media. This is for example the case when diffusion and flow are coupled with physico-chemical reactions at the pore walls (e.g., with adsorption and/or chemical reactions, see discussion in Chap. 4.6). Hence, empirical micro–macro relationships and transport simulations represent different and complementary approaches for studying the properties of porous media, which do not replace each other.

As an introduction to this chapter, recall that the review of tortuosity, performed in this book, basically reveals *two main schools of thinking, where tortuosity is either determined indirectly or,* alternatively, it is *directly* based on morphological descriptors of the underlying 3D microstructure.

Indirect physics-based tortuosity

In a traditional approach, tortuosity is interpreted as the predominant microstructure effect. Following this approach, tortuosity can be determined indirectly, assuming a relatively simple relationship with effective properties (see e.g., Eq. 2.31: $D_{eff}/D_0 = D_{rel} = \varepsilon/\tau_{indir\_diff}^2$). However, the values obtained for indirect tortuosity are consistently higher than what would be expected from geometric considerations of e.g., streamlines. Hence, indirect tortuosity clearly overestimates the lengths of (shortest) transport pathways, which indicates that it includes other microstructure effects such as the limiting influence from narrow bottlenecks. Hence, the indirect tortuosity can be interpreted as bulk microstructure resistance,—normalized by the pore volume. Thus, this information is often used as valuable input for macro-homogeneous models, which intend to describe the bulk microstructure resistance.

Direct geometric tortuosity

An alternative 'geometric' school of thinking is focusing on geometric and mixed tortuosities, which provide reliable estimations of the lengths of transport paths. In order to establish quantitative relationships between microstructure and effective transport properties, it is then necessary to capture all relevant microstructure characteristics—not only the path length effect that is described with direct geometric or mixed tortuosities. This approach thus requires an additional effort, e.g., for determining constrictivity and eventually also hydraulic radius. The quantitative micro–macro relationships obtained by this approach provide a deeper understanding of the relevant morphological effects, which represents the basis for a purposeful microstructure optimization and materials design.

It must be emphasized that the geometric school of thinking profits a lot from the recent progress in 3D image analysis, stochastic geometry and virtual materials testing. These new options for 3D analysis are applied with the aim to establish quantitative micro–macro relationships for porous media. In the following sections selected results of such investigations are summarized first for conductivity and diffusivity, and subsequently also for flow and permeability.

5.2 Quantitative Micro–Macro Relationships for the Prediction of Conductivity and Diffusivity

In this section, micro–macro relationships describing the limiting effects of microstructure on conductivity and diffusivity are reviewed. Thereby, we typically consider transports in porous media like gas diffusion or liquid conduction. However, it must be emphasized that transport in solid phases of a composite material (e.g., ionic or electric conduction in a cermet electrode) is suspended, in principle, to the same microstructure limitations as the conductive or diffusive transport in porous media and can therefore be described with the same morphological characteristics and mathematical equations (see discussion in Chap. 4.5.2).

The progress made in the investigation of micro–macro relationships for conductivity and diffusivity is schematically illustrated in Fig. 5.1. In the following description we follow step by step this illustration from bottom to top.

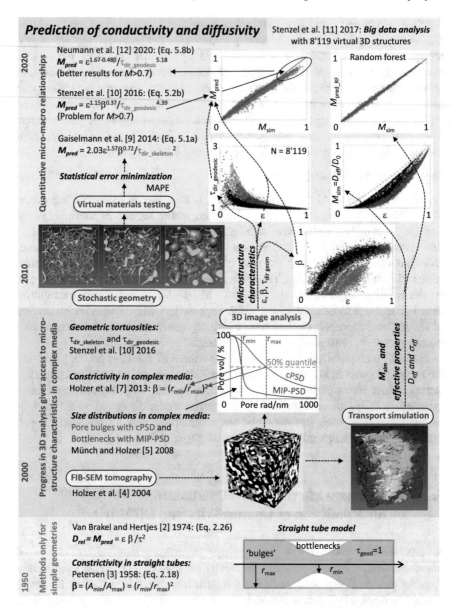

Fig. 5.1 Illustration of progress in microstructure characterization and virtual materials testing. It shows how the evolving 3D methods help to improve quantitative micro–macro relationships, which nowadays enable reliable predictions of effective diffusivity and conductivity even for materials with complex microstructures

Already a long time ago it was recognized that transport in porous media is limited not only by the lengths of tortuous pathways but also by narrow bottlenecks (see e.g., Owen [1]). In 1974, van Brakel and Hertjes [2] thus postulated a micro–macro relationship for conductivity and diffusivity, which includes constrictivity (β) as well as tortuosity (Eq. 2.33: $D_{rel} = \varepsilon\beta/\tau^2$). Unfortunately, at that time, constrictivity as well as direct geometric or mixed tortuosities could not be determined for complex microstructures. Nevertheless, for the simple case of straight tubes ($\tau_{dir\_geometric} = 1$) with varying cross-sections, it was shown by Petersen in 1958 [3] that the retarding impact of bottlenecks can be described by the ratio of the constricted cross-sectional area (A_{min}) over the 'bulged' cross-sectional area (A_{max}). This simple pipe-flow model, which is illustrated at the bottom of Fig. 5.1, led to the definition of constrictivity according to Eq. 2.18 ($\beta = r_{min}^2/r_{max}^2$). However, in the last century the practical relevance of all theories dealing with resistive effects from bottlenecks (constrictivity) and/or path lengths (geometric tortuosity) was strongly limited, since there were no suitable 3D methods available for a quantitative morphological characterization.

With the introduction of FIB-tomography in 2004 [4], 3D imaging of porous media at sub-μm scales became possible. As a next step, suitable tools for quantitative 3D image analysis were required. Two methods to quantify the size distributions of pore bulges and bottlenecks in complex disordered microstructures were introduced by Münch and Holzer [5]. Thereby the continuous pore size distribution (cPSD) was used to characterize the size distribution of pore bulges. Note that the cPSD uses the concept of granulometry functions, which were introduced in [6]. Going beyond granulometry functions, a method for a geometry-based 3D simulation of mercury intrusion porosimetry (MIP) was introduced in [5]. The MIP-PSD (sometimes also called 'porosimetry') reveals the size distribution of bottlenecks. Typical examples of cPSD and MIP-PSD curves are shown in Fig. 5.1.

It was then recognized by Holzer et al. [7] that the 50% quantiles (i.e., r_{50}) of these two pore size distribution curves can be considered as mean effective sizes for bulges ($r_{50\_cPSD} = r_{max}$) and for bottlenecks ($r_{50\_MIP\_PSD} = r_{min}$), which can be substituted in Eq. 2.18 ($\beta = r_{min}^2/r_{max}^2$). In this way, a quantitative method based on 3D analysis was found for the characterization of constrictivity, which also works for materials with complex microstructures. A formal definition of constrictivity in the framework of stochastic geometry was recently provided in [8].

Using experimental data for determining effective properties, as well as constrictivity and geometric tortuosity from 3D analysis, it was soon found that van Brakels equation (Eq. 2.33: $D_{rel} = \varepsilon\beta/\tau^2$) is not very precise in predicting the effective diffusivity. This finding led to the question, which type of equation must be used to describe the relationship between microstructure characteristics and effective diffusivity (D_{eff}) or effective conductivity (σ_{eff}), respectively. In a series of studies [9–12], the following equations were considered as possible candidates:

$$\sigma_{rel}; D_{rel} = M = d\varepsilon^a\beta^b/\tau^c \tag{5.1}$$

$$\sigma_{rel}; D_{rel} = M = \varepsilon^a \beta^b / \tau^c \qquad (5.2)$$

$$\sigma_{rel}; D_{rel} = M = d\varepsilon^a \qquad (5.3)$$

$$\sigma_{rel}; D_{rel} = M = d\varepsilon^a / \tau^c \qquad (5.4)$$

$$\sigma_{rel}; D_{rel} = M = d\varepsilon^a \beta^b \qquad (5.5)$$

$$\sigma_{rel}; D_{rel} = M = d\varepsilon^a \beta^b / \tau^2 \qquad (5.6)$$

$$\sigma_{rel}; D_{rel} = M = \varepsilon^a \beta^b / \tau^2 \qquad (5.7)$$

$$\sigma_{rel}; D_{rel} = M = \varepsilon^{a1-a2\beta} / \tau^c \qquad (5.8)$$

$$\sigma_{rel}; D_{rel} = M = \varepsilon^a \qquad \text{(Archie's law 2.23)}$$

$$\sigma_{rel}; D_{rel} = M = \varepsilon / \tau^2 \qquad (2.31)$$

$$\sigma_{rel}; D_{rel} = M = \varepsilon \beta / \tau^2 \qquad (2.33)$$

According to the 'geometric' school of thinking, *in all these equations τ is thought as a direct geometric tortuosity (τ_{dir_geom},* either geodesic, medial axis or skeleton tortuosity). The prediction power of these equations was investigated thoroughly through a statistical approach of error minimization in [9–12]. For this purpose, models from stochastic geometry were used to generate a large number of 3D microstructures with varying characteristics and effective properties. 3D image analysis was used to compute the microstructure characteristics ($\varepsilon, \beta, \tau_{dir\_geom}, r_{min}, r_{max}$). Numerical transport simulation was exploited to determine effective diffusivity and/ or conductivity (D_{rel}, σ_{rel}) for each virtual 3D microstructure generated by stochastic models. The unknown exponents (a, b, c, d) in the above-mentioned equations were then determined by means of error minimization. As a quality criterion for the predictive capabilities of the above-mentioned equations, the mean absolute percentage error (MAPE) was used in [9], where

$$MAPE\left(M_{sim}, M_{predict}\right) = \frac{1}{n} \sum_{i=1}^{n} \frac{\left|M_{sim,i} - M_{predict,i}\right|}{M_{sim,i}} \cdot 100\% \qquad (5.9)$$

Thereby, M (microstructure-factor) stands for the relative properties (with respect to D_{rel} or σ_{rel}), which were determined in two ways: a) either by numerical simulation (e.g., $M_{sim} = D_{eff\_sim}/D_0$), or b) by substituting the values obtained from 3D image analysis for the microstructure characteristics (ε, β, τ_{dir_geom}) into the equation under consideration (i.e., M_{pred}, for example, $M_{pred} = d\varepsilon^a\beta^b/\tau^c$ in Eq. 5.1). The exponents a, b, c, d were then fitted for each equation (see list of equations above) in order to minimize the corresponding MAPE (i.e., the absolute value of the difference $M_{sim} - M_{pred}$). For the most relevant equations, the values of the fitted pre-factor and exponents as well as the prediction errors (MAPE) are summarized in Table 5.1.

In a first publication of this series of investigations (Gaiselmann et al. [9]), it was shown that for *the traditional equations* (i.e., Eqs. 2.31, 2.32, 2.33 without any fitting of exponents) the prediction power becomes significantly *better when constrictivity is considered* as a relevant microstructure effect, in addition to geometric tortuosity and porosity (cf. Eq. 2.31, where MAPE = 625% versus Eq. 2.33, where MAPE = 37%). Note that the prediction power can be further improved when using equations with fitted pre-factor and exponents (Eqs. 5.1–5.7). From all the equations under consideration, Eq. 5.1 (and Eq. 5.2) revealed the best results with a MAPE of 25% (and 28%, respectively). Then Eq. 5.1 reads as

$$M_{pred} = 2.35\,\varepsilon^{1.57}\beta^{0.71}/\tau_{dir\_skeleton}{}^{2.3}. \tag{5.1a}$$

Again, equations including constrictivity in addition to geometric tortuosity reveal the best predictions. These findings underline the importance of the bottleneck effect. It must be emphasized that the skeleton tortuosity ($\tau_{dir\_skeleton}$) was used throughout [9]. However, other types of geometric tortuosity should be tested as well.

The study presented by Stenzel et al. [10] is based on the same methodologies as in [9]. However, the results of [10] extend the investigations of [9] in the sense that the *prediction power of these equations was compared for different types of geometric tortuosity*. It turned out that the prediction power with *geodesic tortuosity* ($\tau_{dir\_geodesic}$) *is better than with skeleton tortuosity*. For example, the prediction error (MAPE) for Eq. 5.2 improved from 28% with $\tau_{dir\_skeleton}$ to 19% with $\tau_{dir\_geodesic}$. The prediction formula then takes the form

$$M_{pred} = \varepsilon^{1.15}\beta^{0.37}/\tau_{dir\_geodesic}{}^{4.39}. \tag{5.2b}$$

Moreover, *modified definitions of constrictivity* were investigated in [10]. They are based on considerations that the relevant sizes of bulges (r_{max}) and bottlenecks (r_{min}) do not necessarily correspond to the 50% quantiles (r_{50}) of the *cPSD* and *MIP-PSD* curves, respectively. *Other quantiles* (r_0, r_{10}, r_{25}, r_{50}, r_{75}, r_{90}) *from cPSD and MIP-PSD curves were thus considered* as possible candidates for characteristic sizes of bottlenecks (r_{min}) and bulges (r_{max}). By varying the combination of these candidates, Stenzel et al. [10] obtained 35 different r_{min}/r_{max}-ratios as possible definitions for constrictivity (e.g., $\beta = r_{25\_MIP\text{-}PSD}{}^2/r_{75\_cPSD}{}^2$). The impact of these different definitions on the prediction power was then tested for five equations (Eqs. 2.31, 2.33, 5.1, 5.2, 5.7) using the MAPE as a quality criterion. In short, it turned out that *the initial definition of constrictivity* (i.e., $\beta = r_{50\_MIP\text{-}PSD}{}^2/r_{50\_cPSD}{}^2$) from Holzer et al. [7] *gives the most reliable and most precise predictions for all equations*. It is thus proposed to maintain the initial definition of constrictivity.

Table 5.1 Summary of the most important micro–macro relationships (for conductivity and diffusivity) obtained by virtual materials testing, including fitted pre-factor (d) and exponents (a, b, c) for porosity (ε), constrictivity (β) and tortuosity ($\tau_{dir\_geometric}$) and the corresponding mean absolute percentage error (MAPE) as quality criterion

Authors [Ref]	No. of 3D models	Eq. No.	Tortuosity type	d	a (ε^a)	b (β^b)	c (τ^c)	MAPE %	Comments
Gaiselmann et al. [9]	105	2.31	$\tau_{dir\_skeleton}$	–	1	–	2	625	No constrictivity!
Gaiselmann et al. [9]	105	2.33	$\tau_{dir\_skeleton}$	–	1	1	2	37	
Gaiselmann et al. [9]	**105**	**5.1 (a)**	$\tau_{dir\_skeleton}$	**2.346**	**1.569**	**0.709**	**2.298**	**25**	
Gaiselmann et al. [9]	105	5.2 (a)	$\tau_{dir\_skeleton}$	–	1.327	0.689	1.478	28	
Stenzel et al. [10]	43	5.1 (b)	$\tau_{dir\_geodesic}$	0.88	1.06	0.36	4.35	19	
Stenzel et al. [10]	43	5.2 (b)	$\tau_{dir\_geodesic}$	–	1.15	0.37	4.39	19	
Stenzel et al. [11]	**8119**	**5.2 (b)**	$\tau_{dir\_geodesic}$	–	**1.15**	**0.37**	**4.39**	**13.6**	**Not precise for M > 0.7**
Stenzel et al. [11]	8119	Neural network	$\tau_{dir\_geodesic}$	–	–	–	–	8.9	
Stenzel et al. [11]	8119	Random forest	$\tau_{dir\_geodesic}$	–	–	–	–	8.5	
Neumann et al. [12]	**8119**	**5.8 (b)**	$\tau_{dir\_geodesic}$	–	**1.67–0.48β**	–	**5.18**	**18.3**	**Precise for M > 0.7**
								10.3 for M > 0.05	Large errors for M < 0.05

The equations marked in bold emphasis represent the favorites with the highest prediction power. Note that skeleton tortuosity is considered in [9] and geodesic tortuosity in [11]. Also note that quantitative image analysis is always performed only for the contiguous (connected) portion of the pore phase (i.e., ε_{eff}). For example, trapped pores are excluded from the analysis of ε, β and τ.

Up to this stage, the statistical analyses included only a moderate number of 105 [9] and 43 [10] different 3D microstructure models, respectively. In an extensive simulation study by Stenzel et al. [11] the *number of virtual 3D models was increased to 8119*. With this *big data approach*, it was confirmed that effective diffusivity and conductivity are well predicted by Eq. 5.2b. The corresponding MAPE decreased further to 13.6% due to the better statistical data basis. However, it also became clear that M_{pred} and associated effective properties are underestimated by Eq. 5.2b for highly porous materials (i.e., for microstructures with $M > 0.7$). Better predictions, in particular for $M > 0.7$, could be achieved with methods from machine learning, namely random forests, and neural networks, which reveal MAPEs of 8.5% and 8.9%, respectively. However, these tools from machine learning generally do not provide a clear physical interpretation of the microstructure influence on effective transport properties. Therefore, random forests and neural networks could not be used as a basis for microstructure optimization.

In a recent paper by Neumann et al. [12], it was recognized that the problem of Eq. 5.2b for microstructures with $M > 0.7$ originates from an overestimation of the bottleneck-effect at high porosities (particularly in the limit when ε tends to 1). A modified equation (Eq. 5.8) was thus proposed, where constrictivity appears in the exponent of porosity. In this way, constrictivity acts as a correction factor for the effective pore volume, but the effect of β becomes negligible when porosity is close to 1. The 8119 virtual 3D microstructures from [11] were then used as a basis for testing the prediction power of Eq. 5.8b in [12]. Overall, this resulted in a MAPE of 18.3%, which is—when taking all structures into account—not better than the MAPE of Eq. 5.2b obtained in [11]. The prediction formula derived in [12] reads as follows:

$$M_{pred} = \varepsilon^{1.67-0.48\beta} / \tau_{dir\_geodesic}^{5.18}. \qquad (5.8b)$$

However, Eq. 5.8b has the advantage that - in contrast to Eq. 5.2b—it is consistent with theoretical results in the dilute limit, i.e., in the case when the obstacles of the transport process vanish. For materials exhibiting high porosities with $M > 0.7$, the predictions obtained by means of Eq. 5.8b are much better than those obtained by Eq. 5.2b. Nevertheless, for low porosity materials with $M < 0.05$, the predictions by Eq. 5.8b are even worse. Therefore, when considering only structures with $M > 0.05$, the MAPE for Eq. 5.8b improves significantly to 10.3%. Hence, the higher the M-value, the better the prediction power of Eq. 5.8b.

From the numerous equations that were evaluated statistically, the following *three favorite equations* remain (marked with bold emphasis in Table 5.1):

- Eq. 5.1a from [9] gives the best predictions when skeleton tortuosity is used.
- Eq. 5.2b from [10] gives the best results with geodesic tortuosity, but only for microstructures with $M < 0.7$.

- Eq. 5.8b from [12] also gives good results with geodesic tortuosity. In particular, this equation should be used for highly porous materials with M > 0.7 (but not for low porosity materials with $M < 0.05$).

The prediction power of these equations was also validated experimentally for different porous materials by means of tomography (FIB-SEM and μCT). The validation was done by comparing the predicted properties (M_{pred}) from 3D image analysis either with results from experimental characterization (M_{exp}) or from simulations (M_{sim}) or both. In this way, it has been shown in [9, 10, 12] that the equations considered in these papers give good results for SOFC cermet electrodes, including gas diffusivity in the pores as well as electrical conductivities of the solid phases.

Furthermore, Eq. 5.2b *was experimentally validated* for very different types of microstructures, such as sintered ceramic membranes [13], fibrous GDL in PEM fuel cells [14] and even for open cellular materials (unpublished data). These results confirm that the established micro–macro relationships are rather general in the sense that they are *capable to predict effective properties for a wide spectrum of microstructures*, the morphologies of which differ significantly from those used for deriving the microstructure-property relationships. This generality may be surprising, considering the fact that the predictions are based on three volume-averaged parameters (ε, β, $\tau_{dir\_geometric}$) only. However, these findings also indicate that these three parameters indeed capture all microstructure effects, which are relevant for conduction and diffusion, to a large extent. Exceptions are discussed in [15].

Due to the progress in microstructure characterization as well as in mathematical and numerical 3D modeling, many researchers are now considering the distinct morphological limitations that can be described with geometric tortuosity, constrictivity and phase volume fractions. Hence, the *geometric school of thinking is permanently expanding*. For example, quantitative relationships between ε–β–$\tau_{dir\_geometric}$ and effective transport properties are the basis for recent investigations of Li-ion batteries [16], polymer films [17], tight gas reservoirs [18], sandstones [19, 20], packed beds in oil combustion [21], biomaterials/bone tissue [22] and general packings of slightly overlapping spheres [15], see also [23] where large-scale statistical learning has been performed for the prediction of effective diffusivity in porous materials using 90,000 artificially generated microstructures.

5.3 Quantitative Micro–Macro Relationships for the Prediction of Permeability

The evolution of quantitative micro–macro relationships for the prediction of permeability is schematically illustrated in Fig. 5.2. Basically, this evolution can be subdivided into an early period, when methods for 3D-analysis were not yet available, and a recent period after the year 2000, when micro- and nano-tomography, 3D-image processing as well as stochastic 3D modeling became available.

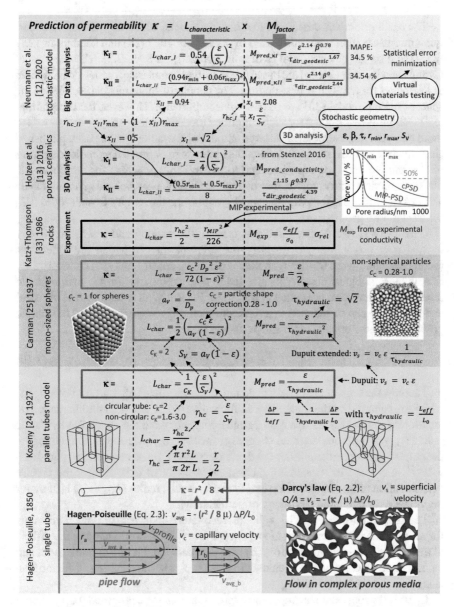

Fig. 5.2 Evolution of quantitative micro–macro relationships for prediction of permeability in porous media

The basic principles describing the limiting effects, which arise from the underlying microstructure, were introduced by Kozeny in 1927 [24] and Carman in 1937 [25] (see also the discussion of hydraulic tortuosity in Sect. 2.2). The theories of Carman and Kozeny [24, 25] are based on simplified geometrical models, which serve as analogues for realistic pore structures. The bundle-of-tubes model [24, 26] and the sphere-packing model [25] are of particular importance. However, the morphological descriptors determined for such simplified models cannot easily be transferred to more complex microstructures. A large amount of literature that has been published since the work of Kozeny and Carman thus intends to improve the prediction power of Carman-Kozeny-type equations and to make them applicable for more realistic materials models with complex microstructures. Early examples are given by Panda and Lake (1994) [27] for poly-dispersed granular media, and by Costa (2006) [28] for fractal pore geometries.

Over the last two decades, the progress in 3D imaging and image processing opened new possibilities to quantify the relevant microstructure characteristics (geometric and mixed tortuosity, bottleneck radius, constrictivity). The availability of new morphological descriptors also led to new expressions for the micro–macro relationships that were presented in literature. Basically, all equations (the classical and new ones) can be reduced to a representation of permeability (κ) by the simple product of characteristic length and M-factor, i.e.,

$$\kappa = L_{char} M. \qquad (5.10)$$

Thereby, the term of characteristic length (L_{char}) accounts for wall friction effects, which are captured by the squared hydraulic radius (r_{hc}^2). Note that the characteristic length term has a major impact on permeability. For example, when the hydraulic radius (r_{hc}) of a porous material changes from 10 to 0.1 μm due to pore clogging, the corresponding permeabilities (κ) decrease by 4 orders of magnitudes, e.g., from $5 \ 10^{-11}$ to $5 \ 10^{-15}$ m^2 (assuming a constant M-factor of 0.5). Hence, the precise determination of the characteristic length and hydraulic radius (L_{char}, r_{hc}) is of major importance for a reliable prediction of permeability.

The second term in Eq. 5.10 is the microstructure M-factor for flow. It accounts for all the other transport limitations, except for wall friction. Thereby, the effects of a) varying pore-volume fractions, b) transport path lengths and c) bottlenecks can be described by dimensionless characteristics (ε, β, τ). The resulting M-factor then takes values between 0 and 1.

In the following sections, important expressions that were proposed for prediction of permeability are briefly reviewed in chronological order. The evolution of these expressions is also visualized in Fig. 5.2 from bottom to top. For better comparison, all expressions are reformulated such that the two terms for characteristic length and M-factor are clearly distinguished.

5.3.1 Bundle of Tubes Model

Kozeny's description of flow [24] from 1927 is based on the consideration of a bundle of tubes (see Chap. 2.2). Flow in a single, straight tube can be described by the Hagen-Poiseuille law (Eq. 2.3). From comparison with Darcy's law (Eq. 2.2) for porous media, it follows that 'permeability' of a single pipe depends only on the radius, i.e.,

$$\kappa_{pipe-flow} = L_{char} = r^2/8. \tag{5.11}$$

Kozeny [24] then introduced a general definition for the hydraulic radius by

$$r_{hc} = \frac{Vol\ open\ to\ flow}{wetted\ surface} = \frac{\pi r^2 L}{\pi 2r L} = \frac{r}{2}, \tag{5.12}$$

which is the ratio of the tube volume open to flow over the surface area that is wetted by the fluid. The combination of Eqs. 5.11 and 5.12 leads to the following 'pipe-flow permeability' given by

$$\kappa_{pipe-flow} = L_{char} = r_{hc}^2/2. \tag{5.13}$$

For porous media consisting of bundles of tubes, the hydraulic radius and associated volume-to-surface ratio from Eq. 5.12 can also be written in terms of porosity over specific surface area per volume (S_V, with units m^2/m^3), i.e.

$$r_{hc} = \frac{\varepsilon}{S_V}. \tag{5.14}$$

The characteristic length term for a bundle of tubes thus becomes

$$L_{char} = \frac{1}{2}\left(\frac{\varepsilon}{S_V}\right)^2 = \frac{1}{c_K}\left(\frac{\varepsilon}{S_V}\right)^2. \tag{5.15}$$

Thereby, the Kozeny factor (c_K) for circular tubes is equal to 2, in accordance with Eq. 5.13. For non-circular tubes the c_K-values vary in the range from 1.6 (for triangular tubes) to 3 (for rectangular tubes with a high aspect ratio).

Furthermore, Kozeny pointed out in [24] that the superficial velocity (v_s) in Darcy's macroscopic description of porous media flow (and thus also for a bundle of tubes) is different from the capillary velocity (v_c) in Poiseuille's microscopic description of flow in a single pipe. According to Dupuit's relation the two velocities are linked by porosity, i.e.,

$$v_s = v_c \varepsilon. \tag{5.16}$$

Kozeny [24] thus introduced porosity as part of the M-factor for flow. In addition, from the comparison of the models for sinusoidal and straight tubes model, it follows that the lengths of the pathways increase from L_0 to L_{eff}. Consequently, the pressure gradient (in Hagen-Poiseuille's and Darcy's law) needs to be corrected accordingly from $\Delta P/L_0$ to $\Delta P/L_{eff}$. In this context, Kozeny [24] introduced the tortuosity concept with the definition of tortuosity ($\tau = L_{eff}/L_0$) in order to describe the path length effect on the pressure gradient. This reads as

$$\frac{\Delta P}{L_{eff}} = \frac{1}{\tau_{hydraulic}} \frac{\Delta P}{L_0}. \tag{5.17}$$

Note that Kozeny's formulation of the M-factor thus includes corrections for porosity and path lengths (tortuosity), which leads to

$$M_{pred} = \frac{\varepsilon}{\tau_{hydraulic}}. \tag{5.18}$$

Combining the M-factor (Eq. 5.18) and the characteristic length term (Eq. 5.15) gives the full Kozeny equation, which describes permeability for a bundle of tubes-model by

$$\kappa_{Kozeny} = \frac{1}{c_K} \left(\frac{\varepsilon}{S_V} \right)^2 \frac{\varepsilon}{\tau_{hydraulic}}, \tag{5.19}$$

with Kozeny's shape factor $c_K = 2$ for circular tubes. It must be noted, that at the time it was not yet possible to distinguish between direct geometric, indirect physics-based or mixed tortuosity. But from the qualitative descriptions, it becomes clear that Kozeny was using the concept of effective path length (such as streamlines), which is best described by mixed tortuosity.

5.3.2 Sphere Packing Model

In Carman's work [25] from 1937, the Kozeny equation (Eq. 5.19) was modified such that it describes flow in a packed bed of spheres (see Sect. 2.2.1.2). For a simplified model with mono-sized spheres, it is straightforward to determine the specific surface area of the spheres (S_P, with P for particle) per solid volume (V_P) of the spheres ($a_V = S_P/V_P$). In order to obtain the specific surface area per total volume of the porous material ($S_V = S_P/V_{tot}$), a correction for the solid volume fraction ($1 - \varepsilon$) is required, i.e.,

$$S_V = a_V(1 - \varepsilon), \tag{5.20}$$

In geometrical models for non-spherical particles, an additional shape correction factor (c_C, the so-called Carman factor) was introduced by Carman [25]. At the same time, Kozeny's correction factor (Eq. 5.15) for tube shape becomes redundant, and thus c_K is replaced with the constant 2. This leads to Carman's characteristic length term

$$L_{char} = \frac{1}{2}\left(\frac{c_C \varepsilon}{a_V(1-\varepsilon)}\right)^2 \tag{5.21}$$

for granular media consisting of mono-sized objects. According to Carman [25], the shape factor (c_C) takes values in the range from 1 (for spheres) down to 0.28 (for platy minerals, mica). The surface-to-volume ratio (a_v) for packing of mono-sized objects (spheres, particles) is often written in terms of the particle diameter ($a_v = 6/D_p$), which then leads to

$$L_{char} = \frac{c_C^2 D_p^2 \varepsilon^2}{72(1-\varepsilon)^2}. \tag{5.22}$$

Kozeny [24] correctly recognized that the pressure gradient in granular media must be corrected for the effect of path length and, therefore, tortuosity was introduced for this correction (see Eq. 5.17). However, it was Carman [25] who realized that the increase of path lengths also has an effect on the computed (superficial) flow velocity, and he therefore extended Dupuit's relationship (cf. Eq. 5.16) for the influence of tortuosity, which leads to

$$v_s = v_c \varepsilon \frac{1}{\tau_{hydraulic}}. \tag{5.15b}$$

In this way, tortuosity was introduced a second time in the M-factor by Carman [25], which leads to

$$M_{pred} = \frac{\varepsilon}{\tau_{hydraulic}^2}. \tag{5.23}$$

Thereby, τ^2 is also called tortuosity factor (T).

Combining the M-factor (Eq. 5.23) with the characteristic length term (Eq. 5.21) gives the full Carman-Kozeny equation

$$\kappa_{C-K} = \frac{1}{2}\left(\frac{c_C \varepsilon}{a_V(1-\varepsilon)}\right)^2 \frac{\varepsilon}{\tau_{hydraulic}^2} = \frac{c_C^2 D_p^2 \varepsilon^2}{72(1-\varepsilon)^2} \frac{\varepsilon}{\tau_{hydraulic}^2}, \tag{5.24}$$

which describes permeability for a packed bed of mono-sized particles. In the appendix of [25], a geometrical model for packed spheres is presented, which enables

one to estimate the length of flow streamlines. Based on these geometric considerations, Carman argued in [25] that the streamline tortuosity in most granular media
must be close to $\sqrt{2}$. Following this argumentation, the value for tortuosity is thus
often fixed at $\tau = \sqrt{2}$. The M-factor for flow then simplifies to $\varepsilon/2$. *In the simplest
form for mono-sized spheres ($c_C = 1$) the Carman-Kozeny equation reduces to*

$$\kappa_{C-K} = \frac{1}{2}\left(\frac{\varepsilon}{S_V}\right)^2 \frac{\varepsilon}{2} = \frac{D_p^2 \varepsilon^2}{72(1-\varepsilon)^2}\frac{\varepsilon}{2}. \tag{5.25}$$

Note that the Carman-Kozeny equation was developed at a time when methods
for 3D imaging and image analysis were not yet available and therefore morphological descriptors were used, which are relatively easy to access (ε, S_V, D_p, τ
$= \sqrt{2}$). Still nowadays, this equation is widely used by the research community.
However, it must be emphasized that the applicability of the Kozeny (Eq. 5.19) and
the Carman-Kozeny (Eq. 5.25) equations are limited to simple microstructures such
as the bundle-of-tubes model and the (mono-sized) packed-spheres model. Already
for relatively small deviations from these idealized geometries (e.g., non-circular
tubes or non-spherical grains) specific correction factors (c_K, c_C) must be fitted,
which introduce considerable uncertainties. *For more complicated materials with
microstructure architectures that are significantly different from sphere or particle
packing, the prediction power of the Carman-Kozeny equation decreases drastically.*

Despite these drawbacks, the Carman-Kozeny equation (Eq. 5.25) is often applied
also for the study of more complex microstructures such as dispersed granular materials and even foams and fibrous materials. However, it was shown by many authors
that the predictions obtained by the Carman-Kozeny equation are highly uncertain
for such complex microstructures [29–32]. Big efforts were undertaken to modify
the Carman-Kozeny equation in order to improve the prediction power also for more
realistic (complex) microstructures, e.g., in [27, 28]. In principle, most of these modifications still use the same relatively simple morphological descriptors (ε, S_V, D_p, τ
$= \sqrt{2}$).

In this context, it is worth to critically consider, which microstructure effects
are reliably captured with the Carman-Kozeny equation, and which are not. From
the above description, it can easily be recognized that the M-factor in the Carman-
Kozeny equation is rather simple. Important microstructure effects resulting from the
variation of path-lengths (constant tortuosity, $\tau = \sqrt{2}$) and bottlenecks (no constrictivity included) are not captured accurately. However, the *strength of the Carman-
Kozeny equation is clearly the description of characteristic length and hydraulic
radius,* respectively, which enables to capture the *wall friction effects* in (mono-sized)
granular media quite well.

5.3.3 Determination of Characteristic Length and M-factor by Laboratory Experiments

Katz and Thompson (1986) [33] presented an experimental solution for measuring characteristic length and M-factor (M_{exp}), which enables to predict permeability of complex porous media. In order to measure the characteristic length, it was proposed to use mercury intrusion porosimetry (MIP). The pore size distribution curve (*MIP-PSD*) typically shows a steep rise, the corresponding radius of which can be interpreted as 'break-through radius' (r_{MIP}). When the pressure is raised to the break-through range, a large portion of the pore space is filled almost instantaneously with liquid mercury. The domain, which is filled with mercury, thus represents a contiguous pore network. Katz and Thompson [33] defined the inflection point of the MIP-PSD curve (convex-concave transition) as break-through radius (r_{MIP}). They argued that r_{MIP} is a characteristic quantity of the pore network, which has a significant influence on flow and permeability, and which can thus be interpreted as an equivalent of the hydraulic radius (r_{hc}). Based on experimental evidence, a constant of 1/226 was determined in [33] as part of this definition of the characteristic length, i.e.,

$$L_{char} = \frac{r_{hc}^2}{2} = \frac{r_{MIP}^2}{226}. \tag{5.26}$$

Moreover, it was argued in [33] that there are additional effects from pore morphology and connectivity, which may have the same limiting influence on flow as they have on electrical conductivity (i.e., non-viscous/non-frictional effects). Consequently, it has been proposed in [33] that the M-factor for flow could be determined based on experimental measurements of effective electrical conductivity (i.e., using porous media saturated with an electrolyte, whereby σ_0 denotes the intrinsic conductivity of the electrolyte). Doing so, an experimental M-factor was obtained, which is defined by

$$M_{exp} = \frac{\sigma_{eff}}{\sigma_0} = \sigma_{rel}. \tag{5.27}$$

Katz and Thompson [33] thus proposed to predict permeability from L_{char} and M_{exp}, using the relationship

$$\kappa_{Katz-Thompson} = L_{char} \frac{\sigma_{eff}}{\sigma_0} = \frac{(r_{MIP})^2}{226} \frac{\sigma_{eff}}{\sigma_0}. \tag{5.28}$$

Note that both characteristics, L_{char} and M_{exp}, are easily accessible with standard experimental methods.

5.3.4   Determination of Characteristic Length and M-factor by 3D Image Analysis

With the *advent of 3D imaging at sub-*μ*m resolution* (e.g., by FIB-SEM tomography in 2004 [4]), it became possible to *quantify specific morphological characteristics in complex microstructures*. Hence, *direct geometric and mixed tortuosities* are nowadays accessible from 3D image analysis and can be used to describe the effects resulting from variations of path lengths. Similarly, *constrictivity (β)* is accessible and can be used to describe the bottleneck effect. Based on these characteristics (ε, β, τ), new expressions for the relationship between microstructure characteristics and conductivity/diffusivity could be established. As described in Sect. 5.2, $M_{pred\_conductivity}$ was determined by Stenzel et al. [10, 11] by using modern methods of stochastic geometry, virtual materials testing and statistical error minimization (compare Eq. 5.2b: $M_{pred} = \varepsilon^a \, \beta^b / \tau_{geodesic}^{\,c}$, with $a = 1.15$, $b = 0.37$, $c = 4.39$).

In analogy to the paper of Katz and Thompson [33] reviewed in Sect. 5.3.3, Holzer, et al. [13] argued that the M-factor of conductivity from Stenzel et al. [10] (Eq. 5.2b) can be used as a first approximation for the M-factor of flow. In addition, for the effective lengths term, two different definitions for hydraulic radius (r_{hc_I}, r_{hc_II}) were proposed in [13]. Note that the two different definitions for hydraulic radius led to two different equations for the prediction of permeability (κ_{pred_I}, κ_{pred_II}).

The first approach presented in [13] uses the classical definition of the hydraulic radius (i.e., the ratio of porosity over specific surface area per volume). However, in contrast to the initial Carman-Kozeny approach, specific surface area (S_V) is not determined from the characteristic sphere diameter (D_p), but it is determined directly from the complex microstructures using 3D image analysis, i.e.,

$$r_{hc\_I} = x_I \frac{\varepsilon}{S_V}, \tag{5.29}$$

where X_I is a fitting parameter (Note: X_I is treated as a constant that is independent from pore morphology). Based on detailed investigations of sintered porous ceramics in [13], the effective properties (κ, σ_{eff}) were determined by experiments (κ_{exp}), pore scale simulation (κ_{sim}) as well as 3D imaging/image analysis (κ_{pred_I}, S_V, β, ε, $\tau_{dir\_geodesic}$). By error minimization ($\kappa_{pred\_I} - \kappa_{sim}$ and $\kappa_{pred\_I} - \kappa_{exp}$) a value of $X_I = \sqrt{2}$ was estimated. This leads to

$$L_{char\_I} = \frac{r_{hc}^2}{8} = \frac{1}{4}\left(\frac{\varepsilon}{S_V}\right)^2 \tag{5.30}$$

as a description of the characteristic length. Permeability (κ_{pred_I}) can thus be obtained from the combination of L_{char_I} (Eq. 5.30) with the M-factor for conductivity from Stenzel et al. [10] (Eq. 5.2b) by

$$\kappa_{pred\_I} = \frac{1}{4}\left(\frac{\varepsilon}{S_V}\right)^2 \frac{\varepsilon^{1.15}\beta^{0.37}}{\tau_{dir\_geodesic}^{4.39}}. \tag{5.31}$$

For the second approach considered in [13], it was argued that the hydraulic radius can also be defined as convex combination of the mean size of bottlenecks (r_{min}) and the mean size of pore bulges (r_{max}), i.e.,

$$r_{hc\_II} = x_{II}r_{min} + (1 - x_{II})r_{min}. \tag{5.32}$$

(We refer to Münch and Holzer [5] for the determination of r_{min} and r_{max}, respectively). Using results from 3D image analysis (κ_{pred_II}, r_{min}, r_{max}, β, τ_{dir_geod}), numerical simulation (κ_{sim}) and experimental characterization (κ_{exp}, for validation) as well as applying error minimization, a value of 0.5 was obtained for x_{II}, which leads to

$$L_{char\_II} = \frac{r_{hc}^2}{8} = \frac{(0.5r_{min} + 0.5r_{max})^2}{8}. \tag{5.33}$$

Permeability (κ_{pred_II}) is thus predicted by a combination of L_{char_II} (Eq. 5.33) with the M-factor for conductivity (see Eq. 5.2b). More precisely,

$$\kappa_{pred\_II} = \frac{(0.5r_{min} + 0.5r_{max})^2}{8} \frac{\varepsilon^{1.15}\beta^{0.37}}{\tau_{dir\_geodesic}^{4.39}}. \tag{5.34}$$

Both approaches (κ_{pred_I}, κ_{pred_II}) were tested with fibrous materials of a gas diffusion layer (GDL) in PEM fuel cells [14]. In-situ time-lapse tomography (μ-CT) was used to capture the changing 3D water-distribution upon ongoing imbibition. A good agreement was obtained between the predicted permeabilities (κ_{pred_I}, κ_{pred_II}) based on 3D characterization with simulated permeabilities (κ_{sim}) based on a numerical 3D flow model. Thereby, the predictions obtained by κ_{pred_II} (Eq. 5.34) resulted in smaller differences to κ_{sim}, compared to the predictions by κ_{pred_I} (Eq. 5.31).

Note that the prediction formula for permeability κ_{pred_II} (Eq. 5.34) presented by Holzer et al. [13] is similar to the prediction proposed by Katz and Thompson [33], in the sense that both approaches use MIP-PSDs (r_{min} and r_{MIP}, respectively) for determining the hydraulic radius (cf. Eqs. 5.26 and 5.32). Furthermore, in both approaches, the M-factor is determined from effective/relative conductivity (cf. Eqs. 5.2b and 5.27). The difference is that physical experiments are used by Katz and Thompson [33], while the approach by Holzer et a l. [13] is based on 3D image analysis.

5.3.5 Determination of Characteristic Length and M-factor by Virtual Materials Testing

Using the same expressions as in Holzer et al. [13] for r_{hc_I} ($= x_I \, \varepsilon/S_V$), r_{hc_II} ($= x_{II} \, r_{min} + (1 - x_{II}) \, r_{max}$) and M_{pred} ($= \varepsilon^a \beta^b / \tau^c$), the corresponding constants and exponents (x_I, x_{II}, a, b, c) were determined recently by means of stochastic geometry and virtual materials testing (see Neumann et al., 2020 [12]). Thereby, the *8119 different 3D microstructures* from Stenzel et al. [11] served as a basis for *big data analysis*. It must be emphasized that in this approach the results obtained with respect of the fitting of constants used in r_{hc} (x_I or x_{II}) and of the exponents used in M_{pred} (a, b, c) are not independent of each other, since the fitting is performed with one simultaneous error minimization procedure. The resulting M-factor is thus specifically fitted for flow and permeability, respectively (i.e., M_{pred_K}). This approach is thus more specific than the permeability predictions of Holzer et al. [13] and Katz and Thompson [33] (see previous sections), where the M-factors are derived from electrical conductivity (i.e., $M_{pred\_conductivity}$).

Using virtual materials testing (i.e., property prediction by 3D analysis and numerical simulation) and error minimization, Neumann et al. [12] obtained a constant of $x_I = 2.08$ for the classical definition of the hydraulic radius ($r_{hc\_I} = x_I \, \varepsilon/S_V$). The resulting description of the characteristic length (L_{char}) is thus very similar to the one from Kozeny [24] for circular pipes with $c_K = 2$, i.e.,

$$L_{char\_I} = \frac{r_{hc}^2}{8} == \frac{1}{8}\left(2.08\frac{\varepsilon}{S_V}\right)^2 = 0.54\left(\frac{\varepsilon}{S_V}\right)^2 = \frac{1}{c_K}\left(\frac{\varepsilon}{S_V}\right)^2. \tag{5.35}$$

Furthermore, for the prediction of permeability considered by Neumann et al. [12], the exponents of M_{pred_KI} are significantly different from those in M_{pred_cond} for conductivity (see Eq. 5.2b, proposed in Stenzel et al. [11]). The fitting revealed a higher exponent for porosity and lower exponent for tortuosity, which leads to

$$M_{pred\_KI} = \frac{\varepsilon^{3.56}\beta^{0.78}}{\tau_{dir\_geodesic}^{1.67}}. \tag{5.36}$$

The full equation for the prediction of permeability (κ_{pred_I}) is then given by

$$\kappa_{pred\_I} = 0.54\left(\frac{\varepsilon}{S_V}\right)^2 \frac{\varepsilon^{3.56}\beta^{0.78}}{\tau_{dir\_geodesic}^{1.67}}. \tag{5.37}$$

For the second case, where the hydraulic radius is determined based on pore size analysis (i.e., $r_{hc\_II} = x_{II} \, r_{min} + (1 - x_{II}) \, r_{max}$), the virtual materials testing revealed a relatively high value of 0.94 for X_{II} (compared to 0.5 that was estimated in [13]). The corresponding characteristic length is then given by

$$L_{char\_II} = \frac{r_{hc}^2}{8} = \frac{(0.94 r_{min} + 0.06 r_{max})^2}{8}. \tag{5.38}$$

This result indicates that the hydraulic radius is almost identical with the mean radius of bottlenecks (r_{min} from MIP-PSD), which is very similar to the definition of the hydraulic radius proposed by Katz and Thomson [33], where $r_{hc} \approx r_{MIP}$.

The M-factor (M_{pred_KII}) for permeability that is obtained from the fitting procedure for κ_{pred_II} is quite different to the M-factors for κ_{pred_I} (see Eq. 5.36) and for conductivity (see Eq. 5.2b), i.e.,

$$M_{pred\_KII} = \frac{\varepsilon^{2.14} \beta^{-0.05}}{\tau_{dir\_geodesic}^{2.44}} \approx \frac{\varepsilon^{2.14}}{\tau_{dir\_geodesic}^{2.44}}. \tag{5.39}$$

For κ_{pred_KII} the exponent for constrictivity (β^b) is slightly below 0, which is counterintuitive from a physical point of view. In [12], it was shown that the statistical error (MAPE) for κ_{pred_KII} is almost identical when comparing Eq. 5.39 with $\beta^{-0.05}$ and Eq. 5.39 without constrictivity (i.e., the case β^0). Hence, constrictivity drops out from Eq. 5.39. This can be explained by the fact that in κ_{pred_KII}, the bottleneck effect is already contained (as r_{min}) in the characteristic length term (L_{char_II}). The equation for κ_{pred_KII} thus becomes

$$\kappa_{pred\_II} = \frac{(0.94 r_{min} + 0.06 r_{max})^2}{8} \frac{\varepsilon^{2.14}}{\tau_{dir\_geodesic}^{2.44}}. \tag{5.40}$$

The statistical analysis performed by Neumann et al. [12] shows that the prediction powers of κ_{pred_I} and κ_{pred_II} are almost identical. The MAPE is 34.5% for both permeability approaches.

5.4 Summary

The *virtual materials testing approach presented by Neumann* et al. [12] is based on a statistical analysis of *more than 8000 different 3D scenarios from stochastic microstructure modeling,* which cover a wide range of microstructures and effective properties. Due to this large data basis, the proposed equations have a rather general character, since they are *capable to predict permeability for various kinds of materials even with very complex microstructures.* For example, μCT-data from cellular, foam like-structures was used in [12] to demonstrate the *high prediction power* of Eqs. 5.37 and 5.40 for materials, which have not been used to fit the parameters in these prediction formulas.

For comparison, the *Kozeny and Carman-Kozeny equations were derived from parallel-tube and packed-spheres models with idealized geometries.* Consequently, the prediction powers of these traditional equations are strongly limited and *not really applicable to more complex microstructures.* An important difference of recently proposed expressions compared to the traditional Carman-Kozeny approach is the introduction of constrictivity in the M-factor. More precisely, it is one of the main shortcomings of the Carman-Kozeny approach, that the *limiting effect resulting from narrow bottlenecks is not properly addressed.*

The equations for the *prediction of permeability (and conductivity)* have also *improved due to a better description of path length effects.* Carman [25] proposed to use a constant value of $\sqrt{2}$ for $\tau_{hydraulic}$. In the approach proposed by Neumann et al. [12], geodesic tortuosity is used. For the 8000 3D microstructures, geodesic tortuosity varies between 1.05 and 2.4 (see Figs. 3.10b and 3.10c). The variation of $\tau_{dir\_geodesic}$ is particularly large (1.2–2.4) for structures with low porosity ($\varepsilon < 0.25$). Hence, neither can tortuosity be considered as a constant ($\sqrt{2}$), as proposed by Carman [25], nor is tortuosity a simple function of porosity, as proposed in widely used tortuosity-porosity relationships (e.g., the Bruggeman relationship, see also discussion in Chap. 3).

For complex microstructures the permeability can only be predicted in a reliable way with suitable descriptions of geometric or mixed tortuosity and other relevant characteristics (effective porosity, constrictivity, hydraulic radius) gained from 3D analysis, see also recent study by Prifling et al. [23], where large-scale statistical learning has been performed for the prediction of permeability in porous materials using 90,000 artificially generated microstructures.

Modern methodologies of 3D analysis open new possibilities for the precise characterization of all transport relevant microstructure characteristics (i.e., ε, β, τ, r_h, r_{min}, r_{max}, S_V). Based on these characteristics, the effective transport properties (conductivity, diffusivity, permeability) can be predicted with a high prediction power.

The most important micro–macro relationships for prediction of effective conductivity and diffusivity are Eq. 5.2b (precise for microstructures with M < 0.7) and Eq. 5.8b (precise for high porosity materials with M > 0.7). For prediction of permeability, Eqs. 5.37 and 5.40 have the highest prediction power.

These *four equations are all based on the direct geodesic tortuosity,* which is independent from the transport process. It is possible, that the *prediction power can be further improved when a mixed physics-based tortuosity is used* (i.e., $\tau_{mixed\_phys\_Vav}$), which combines the transport specific information with the precise geometric analysis of the corresponding path lengths.

It must be emphasized, that the benefit of these quantitative expressions is not only that they can be used to estimate the effective transport properties of porous media. This task can also be fulfilled with dedicated experiments or with numerical simulations. But the micro–macro relationships very much *help to understand, which microstructure feature and which microstructure effect (i.e., pore volume fraction, constrictive bottlenecks, tortuous pathways, viscous drag at pore walls) represents the dominant transport limitation. In this way, these expressions provide important information to materials engineers, which is necessary for a purposeful optimization of the microstructure.*

References

1. J.E. Owen, The resistivity of a fluid-filled porous body. J. Petrol. Technol. **4**, 169 (1952)
2. J. Van Brakel, P.M. Heertjes, Analysis of diffusion in macroporous media in terms of a porosity, a tortuosity and a constrictivity factor. Int. J. Heat Mass Transf. 1093 (1974)
3. E.E. Petersen, Diffusion in a pore of varying cross section. AIChE J. **4**, 343 (1958)
4. L. Holzer, F. Indutnyi, P. Gasser, B. Münch, M. Wegmann, Three-dimensional analysis of porous batio3 ceramics using fib nanotomography. J. Microsc. **216**, 84 (2004)
5. B. Münch, L. Holzer, Contradicting geometrical concepts in pore size analysis attained with electron microscopy and mercury intrusion. J. Am. Ceram. Soc. **91**, 4059 (2008)
6. G. Matheron, *Random Sets and Integral Geometry* (Wiley, New York, 1975)
7. L. Holzer, D. Wiedenmann, B. Münch, L. Keller, M. Prestat, P. Gasser, I. Robertson, B. Grobéty, The influence of constrictivity on the effective transport properties of porous layers in electrolysis and fuel cells. J. Mater. Sci. **48**, 2934 (2013)
8. M. Neumann, C. Hirsch, J. Staněk, V. Beneš, V. Schmidt, Estimation of geodesic tortuosity and constrictivity in stationary random closed sets. Scand. J. Stat. **46**, 848 (2019)
9. G. Gaiselmann, M. Neumann, V. Schmidt, O. Pecho, T. Hocker, L. Holzer, Quantitative relationships between microstructure and effective transport properties based on virtual materials testing. AIChE J. **60**, 1983 (2014)
10. O. Stenzel, O. Pecho, L. Holzer, M. Neumann, V. Schmidt, Predicting effective conductivities based on geometric microstructure characteristics. AIChE J. **62**, 1834 (2016)
11. O. Stenzel, O. Pecho, L. Holzer, M. Neumann, V. Schmidt, Big data for microstructure-property relationships: a case study of predicting effective conductivities. AIChE J. **63**, 4224 (2017)
12. M. Neumann, O. Stenzel, F. Willot, L. Holzer, V. Schmidt, Quantifying the influence of microstructure on effective conductivity and permeability: virtual materials testing. Int. J. Solids Struct. **184**, 211 (2020)
13. L. Holzer et al., Fundamental relationships between 3D pore topology, electrolyte conduction and flow properties: towards knowledge-based design of ceramic diaphragms for sensor applications. Mater. Des. **99**, 314 (2016)
14. L. Holzer, O. Pecho, J. Schumacher, Ph. Marmet, O. Stenzel, F.N. Büchi, A. Lamibrac, B. Münch, Microstructure-property relationships in a gas diffusion layer (GDL) for polymer electrolyte fuel cells, Part I: effect of compression and anisotropy of dry GDL. Electrochim. Acta. **227**, 419 (2017)
15. O. Birkholz, M. Neumann, V. Schmidt, M. Kamlah, Statistical investigation of structural and transport properties of densely-packed assemblies of overlapping spheres using the resistor network method. Powder Technol. **378**, 659 (2021)
16. T. Hamann, L. Zhang, Y. Gong, G. Godbey, J. Gritton, D. McOwen, G. Hitz, and E. Wachsman, The effects of constriction factor and geometric tortuosity on li-ion transport in porous solid-state li-ion electrolytes. Adv. Funct. Mater. **30** (2020)
17. S. Barman, H. Rootzén, D. Bolin, Prediction of diffusive transport through polymer films from characteristics of the pore geometry. AIChE J. **65**, 446 (2019)
18. A.S. Ziarani, R. Aguilera, Pore-throat radius and tortuosity estimation from formation resistivity data for tight-gas sandstone reservoirs. J. Appl. Geophy. **83**, 65 (2012)
19. C.F. Berg, Permeability description by characteristic length, tortuosity, constriction and porosity. Transp. Porous Media **103**, 381 (2014)
20. C.F. Berg, Re-examining Archie's law: conductance description by tortuosity and constriction. Phys. Rev. E. **86**, 046314 (2012)
21. Q. Xu, W. Long, H. Jiang, B. Ma, C. Zan, D. Ma, L. Shi, Quantification of the microstructure, effective hydraulic radius and effective transport properties changed by the coke deposition during the crude oil in-situ combustion. Chem. Eng. J. **331**, 856 (2018)
22. F. Bini, A. Pica, A. Marinozzi, F. Marinozzi, A 3D model of the effect of tortuosity and constrictivity on the diffusion in mineralized collagen fibril. Sci. Rep. **9**, 1 (2019)

23. B. Prifling, M. Röding, P. Townsend, M. Neumann, V. Schmidt, Large-scale statistical learning for mass transport prediction in porous materials using 90,000 artificially generated microstructures. Submitted (2022)
24. J. Kozeny, Über Kapillare Leitung Des Wassers Im Boden. Sitzungsbericht Der Akademie Der Wissenschaften Wien **136**, 271 (1927)
25. P.C. Carman, Fluid flow through granular beds. Chem. Eng. Res. Des. **75**, S32 (1997)
26. A.E. Scheidegger, *The Physics of Flow through Porous Media*, 3rd edn. (Univ. of Toronto Press, Toronto, 1974)
27. M.N. Panda, L.W. Lake, Estimation of single-phase permeability from parameters of particle-size distribution. Am. Assoc. Pet. Geol. Bull. **78**, 1028 (1994)
28. A. Costa, Permeability-porosity relationship: a reexamination of the Kozeny-Carman equation based on a fractal pore-space geometry assumption. Geophys. Res. Lett. **33**, L02318 (2006)
29. D.A. Nield, A. Bejan, *Convection in Porous Media* (Springer, New York, 2013)
30. A.G. Hunt, R.P. Ewing, On the vanishing of solute diffusion in porous media at a threshold moisture content. Soil Sci. Soc. Am. J. **67**, 1701 (2003)
31. Y. Ichikawa, A.P.S. Selvadurai, *Transport Phenomena in Porous Media* (Springer, Berlin, Heidelberg, 2012)
32. S.M.R. Niya, A.P.S. Selvadurai, A statistical correlation between permeability, porosity, tortuosity and conductance. Transp. Porous Media **121**, 741 (2018)
33. A.J. Katz, A.H. Thompson, Quantitative prediction of permeability in porous rock. Phys. Rev. B. **34**, 8179 (1986)

Chapter 6
Summary and Conclusions

Chapter 2

Classical theories of microstructure limitations to transport in porous media (e.g., the Carman-Kozeny equations for flow; Archie's law for conduction) and associated *tortuosity concepts* are reviewed in Chap. 2. These theories were derived a long time ago, when suitable methods for tomography and 3D image analysis were not yet available. The inherent micro–macro relationships are thus based on the consideration of simplified geometry models such as packed beds of mono-sized spheres or parallel tubes. In this way, many aspects of the microstructure could be captured by means of simple morphological descriptors. For example, the wall friction effect and the associated hydraulic radius are described with diameters of spheres or tubes, which are building blocks of simplified microstructure models. Unfortunately, these classical micro–macro relationships with their simplified descriptors cannot easily be transferred to realistic materials with more complex microstructures. Modern adaptations for prediction of effective transport properties in complex microstructures are discussed in Chap. 5.

The *effect of varying path lengths* has been recognized a long time ago as a major microstructure limitation for transport in porous media. Therefore, tortuosity was included as a relevant parameter in the classical theories. However, for practical applications, path lengths and tortuosity are rather complex descriptors, which could not be measured directly from the microstructure until recently. This is one of the reasons why many different definitions, methods and names were introduced in the literature dealing with tortuosity. This multitude of approaches created much confusion, which still nowadays leads to controversial discussions of the topic.

As a countermeasure to this unsatisfactory situation, we propose a *new tortuosity classification scheme*. The classification is based on the selected method, which is used to determine tortuosity (direct versus indirect determination of tortuosities) and

L. Holzer et al., *Tortuosity and Microstructure Effects in Porous Media*,
Springer Series in Materials Science 333,
https://doi.org/10.1007/978-3-031-30477-4_6

on the type of definition (geometric versus physics-based definition of tortuosities). This classification scheme leads to *three main tortuosity categories*:

(a) direct geometric tortuosities
(b) mixed tortuosities
(c) indirect physics-based tortuosities

Based on this classification scheme, we also propose a systematic tortuosity nomenclature, which includes relevant information about the underlying method of determination and details on the geometric or physical definition. The proposed classification scheme and the associated nomenclature are illustrated in Fig. 2.8.

Chapter 3

The relationship between tortuosity and porosity is considered by many authors as a characteristic feature of specific materials classes that can be described with a mathematical expression. To study these relationships, we present an extensive collection of empirical data (tortuosity-porosity couples) from 69 different studies that investigate tortuosity for different materials (appearing in batteries, fuel cells, rocks, packed spheres, fibrous textures, etc.). This collection includes many cases, where different tortuosity types were measured for the same materials. Therefore, these datasets allow a direct comparison of different tortuosity types. This comparison reveals a surprisingly clear picture in the sense that *certain tortuosity types give consistently higher values than others, irrespective of the material under investigation.* With respect to the three main categories of tortuosity, it can be concluded that the measured values for *indirect tortuosities are consistently higher (often >> 2) than those for the mixed and direct tortuosities (often around $\sqrt{2}$ and below).* The observed, systematic order among the various tortuosity types is illustrated in Fig. 3.9.

The empirical data furthermore indicates that the measured *values for tortuosity are more strongly dependent on the type of tortuosity than on the material itself.* These findings underline the importance of carefully selecting a suitable method and to precisely declare the corresponding type of tortuosity with the help of the proposed classification scheme and nomenclature.

The empirical data also shows that *tortuosity-porosity couples do not follow a certain trend in general,* but they are scattering within certain limits. In the dilute limit where porosity approaches the value of 1, tortuosity values asymptotically go to 1 as well, which lowers the upper bound of the scattering field. With decreasing porosity, however, the scattering of tortuosity becomes more pronounced as the upper bound increases. For indirect tortuosities, the upper bound is much higher (up to 20 and more for low porosities < 0.3) compared to direct and mixed tortuosities. Therefore, the scattering of indirect tortuosity is stronger. Based on this observation, it must be concluded that *mathematical expressions for tortuosity-porosity relationships (e.g., the Bruggeman relation) cannot have any universal meaning.* Mathematical tortuosity-porosity formulas can thus only be meaningful when they are derived for a specific tortuosity type and for special microstructure variations, which are discussed in the present paper. Hence, from a generalized point of view, there is much evidence that microstructure characteristics, such as tortuosity, porosity, constrictivity and pore size, can vary independently of each other (within a certain range). In order to describe microstructure effects properly it is therefore necessary to find suitable characterization techniques for all relevant microstructure characteristics.

Chapter 4

An extensive overview of methodologies is given for microstructure characterization in general, and for tortuosity analysis in particular. The workflow for a thorough 3D characterization (see Fig. 4.1) includes several methodologies that are rapidly evolving. Each of these methodologies is reviewed specifically:

(a) tomography
(b) qualitative image processing (3D reconstruction, filtering, and segmentation)
(c) quantitative image analysis (specific algorithms for each tortuosity type)
(d) numerical simulation of transport (conduction, diffusion, flow)
(e) stochastic microstructure modeling and virtual materials testing.

In particular, the different *calculation approaches for the three main tortuosity categories are discussed separately*: The computation of *direct geometric tortuosities* is based on quantitative 3D image analysis. The *indirect physics-based tortuosities* are computed from effective properties, which are determined by numerical transport simulations (or by real laboratory experiments). For *mixed tortuosities*, volume fields obtained by numerical transport simulation are used. The mixed tortuosities are then computed by geometric analysis of these volume fields (i.e., 3D image analysis of streamlines or velocity vectors). Hence, the mixed tortuosities contain information that covers physics-based as well as geometric aspects. In this sense, the *mixed tortuosities can be considered as the most advanced and most relevant descriptors for the path length effect.* For practical help, an *extensive list with available SW packages* and codes for microstructure analysis and modeling is presented (see Table 4.6), with a special emphasize on tortuosity characterization.

Chapter 5

Based on the methodological progress in tomography, 3D image analysis, stochastic microstructure modeling, artificial intelligence and virtual materials testing, new possibilities become available, which allow a thorough characterization of microstructures at different length scales and with different complexities. Investigations using combinations of these modern methodologies provide a *better understanding of the underlying micro–macro relationships.* A prerequisite for these improvements is *better descriptors for the path length effect by means of direct geometric and mixed tortuosities. But also for the bottleneck effect and for the wall friction effect, improved descriptors could be found, such as constrictivity and hydraulic radius based on MIP-PSD and cPSD.* In Chap. 5, it is summarized how these new descriptors were used to establish *new quantitative micro–macro relationships.* Typically, recent approaches are data-driven, and, for this purpose, they involve methods of stochastic geometry, machine learning, virtual materials testing and error minimization. (References to a series of relevant studies in this field are given in Chap. 5).

For conduction and diffusion, the evolution of the most important formulas describing micro–macro relationships is summarized in Fig. 5.1. From the numerous equations that were evaluated, Eq. 5.2b, i.e.,

$$M_{pred} = \varepsilon^{1.15} \beta^{0.37} / \tau_{dir\_geodesic}^{4.39}, \qquad (5.2b)$$

has the highest overall prediction power with a MAPE of 19.06% (Note: the precision power drops for materials with M > 0.7). Thereby, M_{pred} is equivalent to the relative electric conductivity (σ_{ele_rel}) and/or relative diffusivity (D_{rel}). It must be emphasized that for microstructures with a high porosity, more precise predictions are obtained by Eq. 5.8b, i.e.,

$$M_{pred} = \varepsilon^{1.67-0.48\beta} / \tau_{dir\_geodesic}^{5.18}, \qquad (5.8b)$$

with a MAPE$_{total}$ of 18.3% (and a MAPE of 10.3% for microstructures with M > 0.05).

For viscous flow in porous media, the evolution of micro–macro formulas is summarized in Fig. 5.2. Two different expressions for permeability (κ_I, κ_{II}) are derived, namely Eq. 5.37, i.e.,

$$\kappa_I = 0.54 \left(\frac{\varepsilon}{S_V} \right)^2 \frac{\varepsilon^{3.56} \beta^{0.78}}{\tau_{dir\_geodesic}^{1.67}}, \qquad (5.37)$$

with a MAPE of 34.5%, and Eq. 5.40, i.e.,

$$\kappa_{II} = \frac{(0.94 r_{min} + 0.06 r_{max})^2}{8} \frac{\varepsilon^{2.14}}{\tau_{dir\_geodesic}^{2.44}}, \qquad (5.40)$$

with a MAPE of 34.54%. Thus, both formulas have almost the same prediction powers. Moreover, compared to classical theories such as the Carman-Kozeny equation for flow and Archie's low for conduction, these new micro–macro relationships have a much higher prediction power. In particular, they can also be used for complex disordered microstructures, where the classical theories mentioned above are not applicable. These improvements are mainly due to the progress of recent 3D methodologies, which provide better morphological descriptors.

Interpretation of the three main tortuosity categories

Nowadays many different possibilities are available for the characterization of tortuosity. In this context the findings from the *review of empirical data* (Chap. 3) must be kept in mind. Thereby *a consistent pattern is observed, which indicates that the measured values of tortuosity are more strongly dependent on the tortuosity type (and on the associated method) than on the material itself.* It is thus important to understand the differences between specific tortuosity types. Which type of tortuosity and which calculation approach one should choose, depends on the information that is required for a specific purpose. The basic arguments for different tortuosity classes can be summarized as follows:

(a) *Indirect physics-based tortuosities* describe *bulk resistances of the microstructure* against specific transport processes. They do not contain strict geometric information and therefore they do not really contribute to a fundamental understanding of the path length effect. Indirect tortuosities are *lumped parameters, which include not only the limiting effect of paths lengths variations but also other microstructure limitations such as the bottleneck effects.* Indirect tortuosities are often used as input for macro-homogeneous models. For this purpose, they are well suited, since they describe the bulk resistive influence of the microstructure.

(b) *Direct geometric tortuosities* are based on *measurements of path lengths* through the 3D microstructure. For materials engineers, the geometric tortuosity reveals morphological information that is relevant for purposeful microstructure optimization. However, microstructure limitations on transport cannot be fully described by the geometric tortuosity alone. In order to understand the relationship between microstructure and effective transport properties, *it is necessary to consider additional microstructure characteristics, such as constrictivity, porosity and hydraulic radius.* In context with new micro–macro formulas (e.g., Eq. 5.2b), the *geodesic tortuosity* turns out to be a suitable geometric descriptor for path-length effects. However, since the geometric tortuosities are *not physics-based*, this leaves some room for further improvement of micro–macro formulas and their prediction power.

(c) *Mixed tortuosities* include the *advantages of both calculation approaches*, in the sense that they contain *true information of the path lengths (i.e., geometric)* and at the same time, they are *specific for the underlying transport process (i.e., physics-based).* Mixed tortuosities thus bear key information that is necessary to understand path length effects of specific transport mechanisms on a fundamental level. It is thus probable that the prediction power of modern micro–macro relationships (such as Eq. 5.2b and Eq. 5.37) can be further improved in future by using mixed tortuosities as descriptors for the paths length effect.

Printed in the United States
by Baker & Taylor Publisher Services